DEVELOPMENT OF EUROPEAN FORESTS UNTIL 2050

A Projection of Forest Resources and Forest Management
in Thirty Countries

EUROPEAN FOREST INSTITUTE RESEARCH REPORT

The Scientific Advisory Board:

PROF. DR GÉRARD BUTTOUD, France
MR PÉTER CSÓKA, Hungary
PROF. DR MAX KROTT, Germany
ASS. PROF. DR ELENA G. KULIKOVA, Russian Federation
PROF. DR JARI KUULUVAINEN, Finland
PROF. DR FRITS MOHREN, The Netherlands, Chairman of the SAB
PROF. DAVIDE PETTENELLA, Italy
DR EDUARDO ROJAS-BRIALES, Spain
PROF. DR HUBERT STERBA, Austria
ASS. PROF. DR MARGARIDA TOMÉ, Portugal

NO. 15

DEVELOPMENT OF EUROPEAN FORESTS UNTIL 2050

A Projection of Forest Resources and Forest Management in Thirty Countries

European Forest Institute Research Report 15

BY

G.J. NABUURS
R. PÄIVINEN
A. PUSSINEN
M.J. SCHELHAAS

BRILL
LEIDEN · BOSTON
2003

The views expressed in this book are those of the authors and do not necessarily correspond to those of the European Forest Institute.

This book is printed on acid-free paper.

Library of Congress Cataloging-in-Publication Data

The Library of Congress Cataloging-in-Publication Data is also available

Die Deutsche Bibliothek - CIP-Einheitsaufnahme

Development of European forests until 2050 (A projection of forest resources and forest management in thirty countries)
/by G.J. Nabuurs ...—Leiden ; Boston ; Köln : Brill, 2002
  (European Forest Institute research report; No. 15)
  ISBN 90–04–12311–3

ISSN 1238-8785
ISBN 90 04 12311 3

© Copyright 2002 by Koninklijke Brill NV, Leiden, The Netherlands

All rights reserved. No part of this publication may be reproduced, translated, stored in a retrieval system, or transmitted in any form or by any means, electronic, mechanical, photocopying, recording or otherwise, without prior written permission from the publisher.

Authorization to photocopy items for internal or personal use is granted by Koninklijke Brill NV provided that the appropriate fees are paid directly to Copyright Clearance Center, 222 Rosewood Drive, Suite 910, Danvers, MA 01923, USA.
Fees are subject to change.

PRINTED IN THE NETHERLANDS

CONTENTS

ACKNOWLEDGEMENTS vii
EXECUTIVE SUMMARY ix

1. INTRODUCTION 1
   1.1. Forest resources and functions in Europe 1
   1.2. Wood supply and demand issues 4
   1.3. Conservation and biodiversity issues 6
   1.4. Environment and carbon sequestration issues 8
   1.5. Employment and rural development issues 9

2. AIM 11

3. INVENTORY DATA AND MODELLING APPROACHES 13
   3.1. Input data 13
   3.2. The age class area matrix approach 15
   3.3. Diameter distribution based approach 17
   3.4. Simple balance approach 21

4. SCENARIOS 23
   4.1. Basic Assumptions made 23
   4.2. Scenarios 25

5. COUNTRY LEVEL RESULTS 27
   5.1. Albania 29
   5.2. Austria 35
   5.3. Belgium 41
   5.4. Bosnia and Herzegovina 47
   5.5. Bulgaria 53
   5.6. Croatia 59
   5.7. Czech Republic 65
   5.8. Denmark 71
   5.9. Finland 77
   5.10. France 83
   5.11. Germany 89
   5.12. Greece 95

5.13.	Hungary	101
5.14.	Ireland	107
5.15.	Italy	115
5.16.	Luxembourg	121
5.17.	Macedonia	127
5.18.	The Netherlands	133
5.19.	Norway	139
5.20.	Poland	145
5.21.	Portugal	151
5.22.	Romania	157
5.23.	Slovak Republic	163
5.24.	Slovenia	169
5.25.	Spain	175
5.26.	Sweden	181
5.27.	Switzerland	187
5.28.	Turkey	193
5.29.	United Kingdom	199
5.30.	Yugoslavia	205

6. EUROPEAN SCALE RESULTS 211
   6.1. European totals 211
   6.2. Problems in finding desired fellings 216
   6.3. Spatial distribution of the results 217

7. DISCUSSION 223
   7.1. Uncertainties 223
      7.1.1. Initial state 223
      7.1.2. Basic assumptions related to simulations .. 226
      7.1.3. Simulated results for increment and mortality ... 227
      7.1.4. Model behaviour and validation 229

8. EUROPE'S FOREST RESOURCE POTENTIAL AND
   POLICY IMPLICATIONS 233
   8.1. Summarizing the results 233
   8.2. Forest resource development in socio-economic and
        policy context 234
   8.3. Shaping the policy options for the future 235
   8.4. What should we know to control the development? . 237

REFERENCES 239

ACKNOWLEDGEMENTS

This study has been made possible through funding by the European Forest Institute, the ALTERRA Institute, the City of Joensuu, the European Union funded project LTEEF-II (ENV4-CT987-0577), the Finnish Ministry of Agriculture and Forestry, and the Dutch Ministry of Agriculture, Nature Management and Fisheries. We are greatly indebted to Prof. Ola Sallnäs and Jocke Fredriksson from the Swedish University of Agricultural Sciences for providing the core of the growth simulator of the EFISCEN model to the European Forest Institute. Furthermore, we want to thank the country data correspondents for providing the national forest inventory data. These correspondents are credited at the beginning of appropriate chapter in the current report. Also, our thanks to Mr Tim Green, who revised the language of this report, and to Mr Ilpo Tuononen and Ms Minna Korhonen of EFI who finalised the manuscript for publishing. Last, we wish to thank those people who supported this study from the very beginning Prof. Birger Solberg, Prof Frits Mohren and Dr. Anton Sepers.

EXECUTIVE SUMMARY

This report presents the results of a simulation study concerning European forest resource development until 2050. Under varying assumptions for demand and management, the development of the forest resource of each of 30 European countries in this study is presented, covering altogether 139 million ha of forests available for wood supply. The presented variables (age class development, fellings, increment, growing stock, natural mortality and share of set-aside area) address production potentials, but can equally be regarded as indirect indicators of biodiversity and carbon sequestration potential.

Due to differences in data availability, three modelling approaches were used for this large-scale forest scenario study: (1) an even-aged age class area matrix modelling approach was applied to 114 million ha; (2) a diameter distribution approach was applied to 20.9 million ha; and (3) a simple balance method was applied to 4 million ha. Generally, single forest types were distinguished down to the province, tree species, owner, site class, and age class levels. The simulations of the alternative scenarios (business-as-usual (BAU), maximum sustainable production (MAX), replication of ETTS-V, and multi-functional (MTF)) should be seen as projections, not as predictions.

Except for the maximum sustainable production scenario, the projections show a continuous build-up of growing stock. Gross annual increment will reach a level of around 4.8 m^3/ha/yr in 2050 providing a total increment varying between 637 million m^3/yr (maximum sustainable production) and 729 million m^3/yr (multi-functional).

If fellings remain at the current total level of just under 400 million m^3/yr (business-as-usual), the average growing stock in Europe will rise from 137 in 1990 to 226 m^3/ha in 2050. This continuous rise of growing stock levels off only slightly due to increasing natural mortality of some 13% of increment in this scenario in 2050.

Foreseen realistic increases in fellings in the multi-functional scenario of some 0.3% per year do not do much to curb this trend of increasing growing stock. If the fellings increase to 450 million m^3/yr by 2030, as foreseen in the MTF scenario, then the average growing stock in Europe for all forests (both managed and set-aside) still rises to 218 m^3/ha in 2050. The assumed increase in area of set-aside from 4 million ha in 1990 to 11.9 million ha in 2050, does not hamper the required fellings increase of the MTF scenario. The total production of dead wood in this scenario amounts to 83 million m^3/yr in 2050.

The projections show a maximum sustainable fellings level of 606 million m^3/yr. This amounts to some 95% of the gross increment, and 100% of the net increment.

Under this level the growing stock already stabilises at around 143 m^3/ha. This maximum sustainable level is determined by increased mortality, decreasing increment through time, and also by management constraints as set in this study.

At first glance the scenarios do not provide any dramatically different outlook for the future of forests in Europe. From the policy formulation perspective, however, the information obtained from the scenarios contains a very important message and the main conclusion of the study: European forests and their potential provide room for policy making to design socially, economically and environmentally balanced policy options.

Within the limits of sustainable forest management there are opportunities, in most of the countries, for increased economic utilisation of cutting possibilities while at the same time ensuring the nature conservation objectives. The increasing resource provides policy makers with a challenge and the luxurious situation to choose between varying combinations of aims for biodiversity values and increased cuttings for rural development, bio-energy, and employment opportunities according to national priorities.

The projected scenarios of European forests pose a fundamental question for policy making in European societies: should business-as-usual policies be preferred by policy makers (and the public) in the future, or is there a need to re-design policy interventions to contribute to the utilisation of these potentials for the welfare of European people and sustainable development? Policy discussions during the 1990s showed that in changing socio-economic and environmental conditions the status-quo policy option may not be the preferred one.

In circumstances where new and innovative policy formulations are called for, a second series of questions about 'what should we know?' will arise. What are the impacts of alternative policy means? How effective and efficient are regulative, financial or informational instruments in support of biodiversity conservation, rural development or sustainable timber supply? And last, what are the direct and indirect effects regionally and/or for future generations? This report provides some answers to the questions asked above.

1. INTRODUCTION

The choices that are made in forest management today will have an impact on that forest resource for at least 100 years to come. Therefore, insight into alternative future developments is of great importance. When wishing to generate that insight through projections, an understanding of the processes that created the current forests is essential. In this chapter we therefore look into trends in current European forests. Then the aim of the study is defined (Chapter 2), after which methods (Chapter 3) and scenarios (Chapter 4) are given. Country level results are given in Chapter 5; European level results are given in Chapter 6. The results are discussed in Chapter 7; and the policy implications are discussed in Chapter 8.

1.1. Forest resources and functions in Europe

The primeval forest that once covered most of the European land area has widely been converted to agricultural and other land use purposes over thousands of years. In the early 19th century, the opposite process started to take place, often to guarantee strategic benefits such as timber for construction, or fuelwood for melting ore (Figure 1.1). Today European forests cover 31% of the land area, i.e. 175.8 million ha, of which 149 million ha are available for wood supply (UN-ECE/FAO 2000)[1].

These afforestations and the regular management of practically all of European forests have resulted in a forest cover with a current average age of 57 years. The average standing volume amounts to 142 m^3/ha and the net annual increment is 4.6 m^3/ha/yr (UN-ECE/FAO 2000). However, these average figures do not characterise Europe's

---

[1] UN-ECE/FAO included 38 European countries excluding the area of the CIS. European forests refers in the current study to the forests of the following 30 countries: Albania, Austria, Belgium, Bosnia and Herzegovina, Bulgaria, Croatia, Czech Republic, Denmark, Finland, France, Germany, Greece, Hungary, Ireland, Italy, Luxembourg, Macedonia, the Netherlands, Norway, Poland, Portugal, Romania, Slovakia, Slovenia, Spain, Sweden, Switzerland, Turkey, United Kingdom and Yugoslavia (see also Figure 2.1).

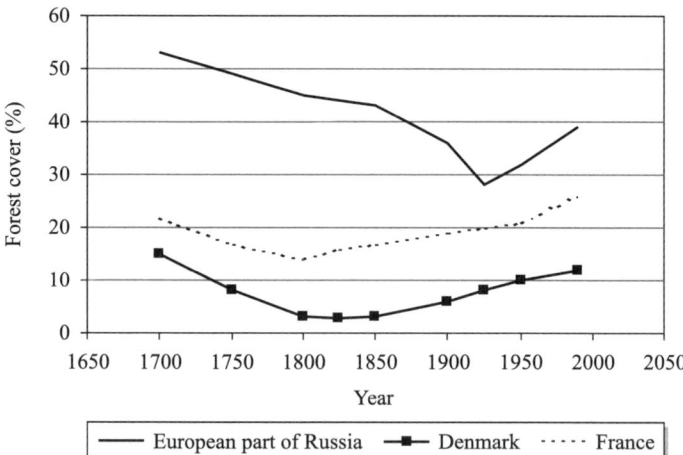

Fig. 1.1. Exemplary historical development of forest cover as a percentage of total land area in some countries since 1700 (Pisarenko et al. 2001, Mather 1990). Primeval forest cover in these countries is estimated to have amounted to 80%, but this is uncertain.

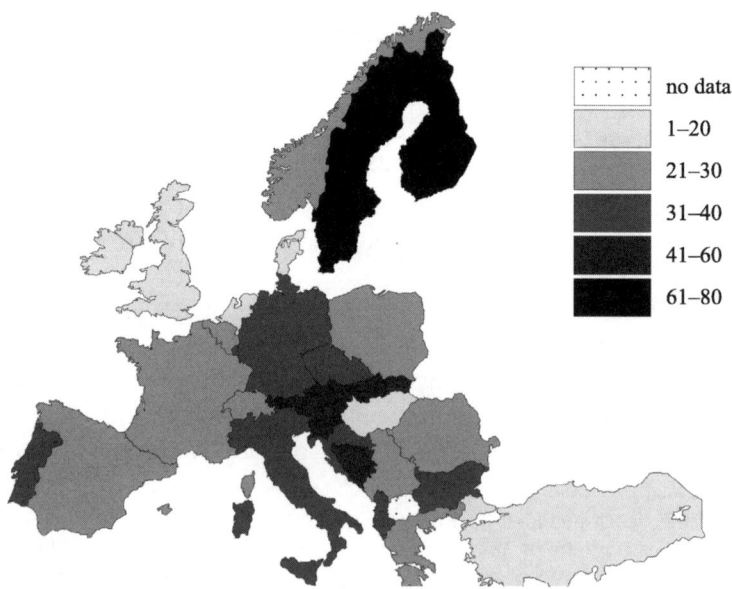

Fig. 1.2. The proportion of the land area covered with forest in each of the European countries under study.

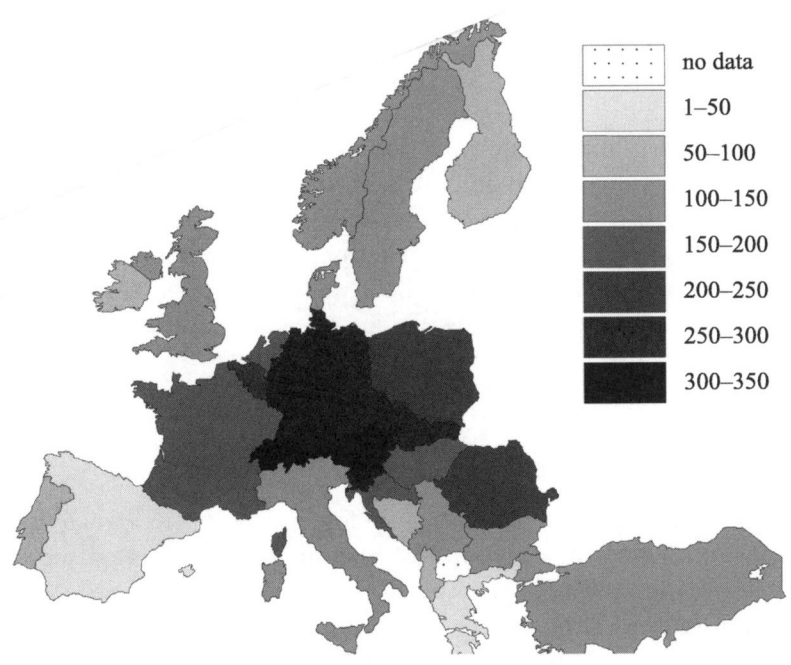

Fig. 1.3. The average growing stock volume (m³/ha) in the European countries under study.

forests so well; there is a huge variety between countries in forest cover (Figure 1.2), species composition, age class distribution, growing stock (Figure 1.3), and aims and culture of forest management. This should always be kept in mind when we talk about 'harmonised European forest resource projections'.

The present net annual increase of forest area is very small, amounting to 0.51 million ha/yr (UN-ECE/FAO 2000), despite an annual deforestation of 0.075 million ha/yr in the EU15 countries (Liski et al. 2000).

European forests fulfil many functions: social functions in providing a living environment and work opportunities, recreation and landscape. The role of the forest is also important in the cultural life, providing aesthetic and spiritual experiences, and conserving the natural ecosystem. They fulfil protective functions for the climate and soils, and they are important for conservation of biodiversity. Furthermore they are a renewable natural resource, producing wood as a raw material, game and berries. All of these functions are fulfilled according to economic forces, but also according to what society and forest policies desire. These phenomena together steer forest management and the everyday choices that a forest owner makes. Some of the functions may be achieved simultaneously in a sustainable way (such as protective and aesthetic values), and some others may conflict with each other to some degree (such as intensive wood production and conservation). Nevertheless, forest management in Europe has aimed at achieving the desired mix of functions in an optimal way. This mix can spatially be achieved from a very local scale (within one stand) up to the European scale.

1.2. Wood supply and demand issues

Figure 1.4 portrays an increasing difference between annual forest growth and fellings in Europe. The increment has increased substantially since 1960 partly as a result of the increased growing stock and area, and improved management, but also possibly as a result of environmental changes (Spiecker et al. 1996). However, it is obvious that the growing stock in Europe is building up quite rapidly. Many reasons prevent the full utilisation, if desired at a large scale, of the timber resources.

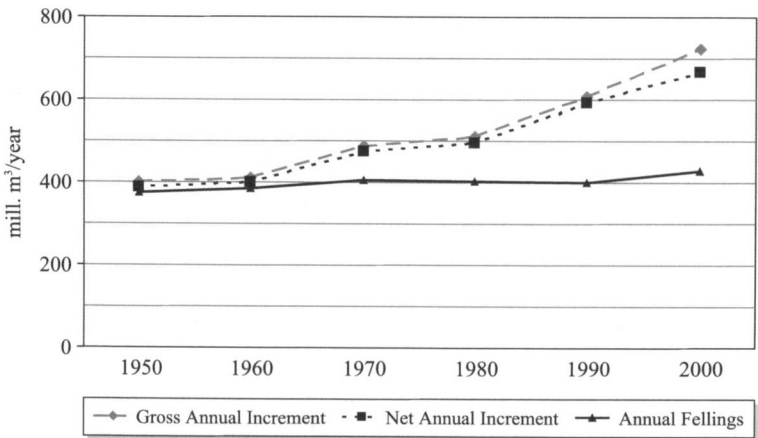

Fig. 1.4. Gross and net increment and annual fellings in the forest area available for wood supply in Europe (excluding CIS). Source: Kuusela 1994 and UN-ECE/FAO 2000.

Figure 1.5 shows that the 30 countries under study meet slightly more than 10% of their roundwood consumption through net import from outside their area. That import includes the trade from CIS countries to the Western European part of the continent, and the import of roundwood from other continents (UN-ECE/FAO 1996, COMTRADE 1998, EFI/WFSE database). Analyses of these trade databases show that most of the trade is within the European continent; very small amounts are traded with other continents. This means that Europe is highly dependent on its own forest resources.

On the supply side a decline of the global forest resource is apparent, which may put more pressure on wood production from European forests in the future. On the other hand, forest ownership is changing. Fewer forest owners in Europe live in the countryside and fewer are dependent on the forest for their income. In some countries (e.g. the Netherlands), forest area is being acquired by nature conservation organisations which, after an initial period in which they remove exotic tree species rapidly, may not be willing to harvest wood.

Ownership changes may also have the opposite effect. Privatisation of forests in Central European Countries and in the newly Independent States with economies in transition has, in some cases, led to sharp increases in the number of forest owners and in the level of harvesting, with a possible continuing financial interest in forests (Csoka 1998).

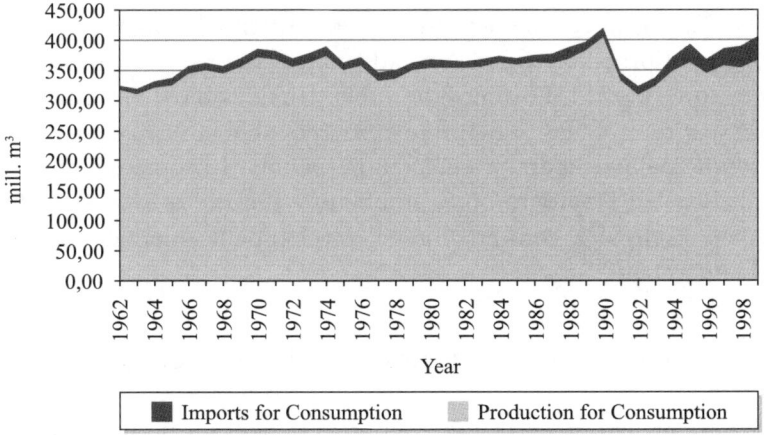

Fig. 1.5. Meeting the apparent consumption (roundwood equivalents) in Europe through domestic production and the trade balance (COMTRADE 1998, EFI/WFSE database).

Furthermore, when the Russian forest sector recovers from its present recession, it could flood the European wood market and wood prices might drop (Anttonen and Petrov 1997). In contrast, increased recycling of abandoned wood products may again reduce the demand for raw material.

Natural disturbances that may increase under increasing growing stock will cause anomalies in the wood supply and may disrupt the wood market for several years. The storms of 1990 and 1999 caused unplanned fellings, which in the last storm were equivalent to half of the total European fellings (Schelhaas et al. 2002, UN-ECE/FAO 2000).

Other changes may occur on the demand side too. Impacts on the demand may result from the European Commission's goal to double the share of renewable energy – partly wood-based – to reach 12% by the year 2010 (European Commission 1997). Studies suggest this may enhance demand by 163 million m^3/yr and raise roundwood prices (Dielen et al. 2000). In addition to the impacts of such a new policy, demand may also increase with increasing GDP. For example, Trømborg et al. (2000) project that between 1994 and 2010, consumption of wood products in Europe will increase annually by between 0.78% and 2.77%, depending on the commodity. For the USA, Haynes et al. (2000) project an overall consumption increase of 0.9% per year until 2050.

The increasing consumer interest in sustainably produced timber may give European wood an advantage over wood from other regions in the world. Furthermore, policy makers have recognised that forests are one of few renewable resources that are able to produce environmentally friendly products in perpetuity. In some countries this has already led to programmes stimulating the use of wood products. The net result of a combination of developments mentioned here is highly uncertain.

1.3. Conservation and biodiversity issues

Forests are generally seen by the public as refuges for nature in Europe, even though the European forests sometimes strongly deviate from the potential natural vegetation (Uuttera et al. 1997). Other reasons in addition to management that have brought about the deviation from potential natural vegetation are environmental pollution and the fragmentation of forests in an agricultural and urbanised landscape.

There are two basic approaches in the conservation of forest biological diversity:

- Set aside forest land from other use to have conservation as the main objective for the area;
- Develop the forest management systems to maintain biological diversity while simultaneously fulfilling other forest functions.

To gain insight into the degree to which the first option has been applied, one possibility is to look at the proportion of the forest not available for wood supply. In Figure 1.6, we can see that in Spain, Sweden and Italy, forest not available for wood supply is close to one-fifth of the total forest area (UN-ECE/FAO 2000). However, that can be due to a combination of conservation and/or economic reasons. These figures, as reported for TBFRA, need further interpretation when compared with other sources. For example, in Parviainen et al. (1999) France, Netherlands and Romania are reported to have 4%, 8%, and 5.5% of the forest area as protected, respectively; but Sweden is reported to have only 3.7% and Italy 6.7% of the forest area as protected.

On the other hand, the managed forest is also reported to be more diverse at the landscape scale, and in many cases the long history of management has created very rich cultural forest types (Kirby and Watkins 1998). In general it is widely agreed that the nature conservation and biodiversity function of European forests is becoming

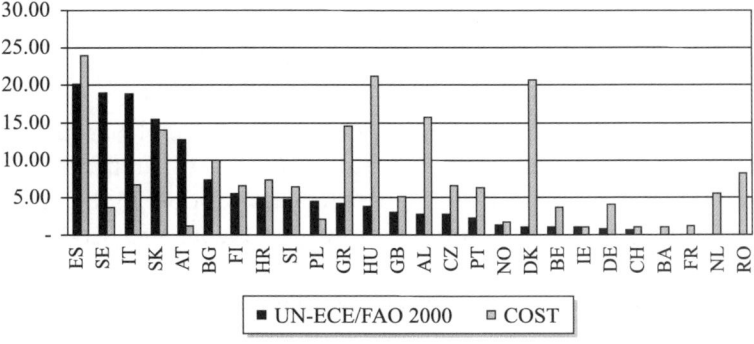

Fig. 1.6. Forest percentage not available for wood supply as reported by two sources (UN-ECE/FAO 2000, Parviainen et al. 1999).

more important (Glück et al. 1999, Farrell et al. 2000). As a result, the perception of forest management and the philosophies underlying it are gradually changing (Kennedy et al. 1998). This gradual change in perception is being strengthened by local problems with monocultures (Klimo et al. 2000).

All over Europe, forest management nowadays aims to enhance the nature conservation value of forests (Pelkonen et al. 1999). Some of the important underlying driving factors are post-modernisation tendencies in societies and the continuing alienation of people from their land base. Such nature-oriented management philosophies are reflected not only in the highly urbanised societies, but also in traditional forest dependent economies. These changes will most likely lead to a major change in the constitution of the European forests, particularly when backed up by a European forest strategy (Fischler 1998) or international conventions such as the Habitat Directive, the EU Biodiversity Strategy and the NATURA 2000 programme. This may result in shifts in the wood-producing regions in Europe and thus in the international trade flows in wood products.

1.4. Environment and carbon sequestration issues

Environmental changes (e.g. climate change) will have implications for the long-term development and existence of the forest resource. Small changes in precipitation pattern, temperature and the occurrence of storms will affect critical processes such as regeneration and seedling establishment, occurrence of insects and fungi, soil organic matter decomposition, nutrient availability, competition between tree species, and stability of mature stands (Kirschbaum et al. 1996, Kellomäki et al. 2000). Climate change will thus affect the role that European forests play in the future terrestrial biospheric carbon balance. European forests are currently estimated to compensate for 11% of Europe's CO_2 emissions from fossil fuel (Nabuurs et al. 1997, Martin et al. 1998, Liski et al. 2000, Valentini et al. 2000), which is almost double the emission reduction that the EU is committed to in the Kyoto Protocol (Watson et al. 2000).

Whatever forest management regime might finally be adopted under the Protocol, measures to continuously increase the standing volume or at least keep it at the present level, seem likely responses. This may generate a new source of finance for forestry. For example, carbon has acquired a monetary value through the introduction of CO_2 emission taxes, estimated at €57/ton CO_2 in Norway. This is comparable to the stumpage value of wood as a raw material (Solberg 1995). Owners may, therefore, start looking at their forest not only as a raw material for industry, but also as a reserve of carbon.

Another environmental issue is the ongoing defoliation of the forests in Europe. In 1999 almost a quarter of all trees inventoried in Europe were classified as moderately to severely damaged, even if the change in the amount of damaged trees has been minimal over the past decade (UN-ECE and European Commission 1999).

1.5. Employment and rural development issues

The share of the forest sector in Gross Domestic Production is generally low in European countries. Even in Finland and Sweden, countries with a high forest cover, the forest sector contributes less than 5% to the GDP.

Since 1987 there has been a downward trend in the number of farms possessing forest (Wermann 1999). This is partly caused by the poor profitability of European forests, which averages a gross income of €65/ha/yr (Pelkonen et al. 1999, Niskanen pers. comm.). It is

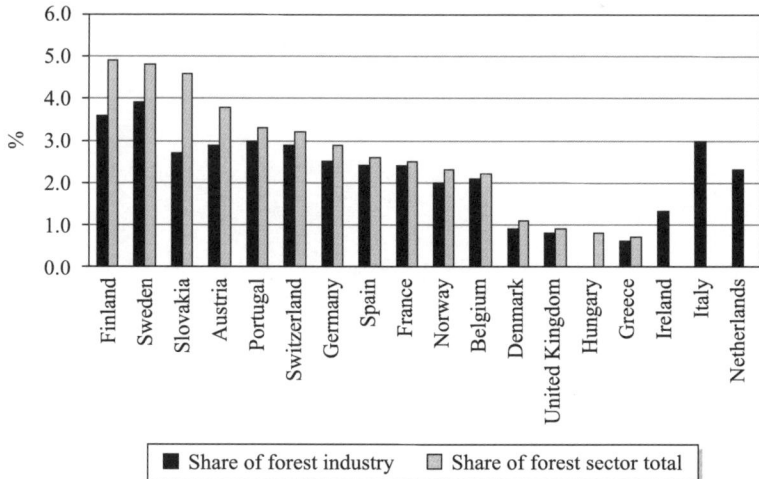

Fig. 1.7. The share of forest sector employment in total national employment.

also partly caused by the changing character of the private ownership: namely, that fewer owners depend on the forest for their income (Peck 1998). In many countries ownership is dominated by the state, but in others, such as Finland and Sweden, a significant proportion of the forest is owned by forest industry. Yet even in those two countries, forestry accounts for less than 5% of national employment (Päivinen et al. 1999a; Figure 1.7).

From this perspective it is understandable that forestry – even if it may have high importance in rural areas in all European countries – is an unimportant issue in the national employment and income policy in most European countries.

On the other hand, forests provide financial benefits to a much wider community than just the forestry sector. Through recreation and tourism the existence of forests provides substantial financial benefits to regional economies, thus keeping the rural areas alive.

2. AIM

The introduction has sketched a European forest resource that fulfils a multitude of demands for a highly urbanised society in Europe. These demands are broadening and constantly changing. European countries are dependent on each other in forest matters, not only through import and export of wood-based products, but also in the context of environmental, conservation and employment issues, as described before.

Forest matters in the European countries will become more and more integrated, especially when the EU will be enlarged with the current pre-accession states, and particularly when backed by a European forest strategy (Fischler 1998). The rapidly changing societal and political demands will have to be met by a forest resource that can only be changed slowly. Therefore, long-term analyses are needed especially where choices that are being made today will have an impact on the forest resource for at least 50–100 years to come. Furthermore, to support the discussion on potential pan-European forest strategies, harmonised scenarios on the possible future of the forests are indispensable.

So far, harmonised dynamic forest resource projections for Europe have only been done once based on inventory data of the early 1980s (Nilsson et al. 1992). With the fast changing political arena and demands from society, the results of that study were becoming outdated in terms of assumed defoliation effects, scenario assumptions, and input inventory data. Therefore, the aim of the present study was to improve the dynamic harmonised approach and to apply it to the European forests in order to create, describe, and provide insight into alternatives for the future of these forests (Figure 2.1).

More specifically, we aimed at harmonised projections of the future development and management options of European forests under increasing and possibly conflicting demands. In that way its results will provide forest policy makers at the national and European level with quantitative insights into the long-term effects of current trends.

The move towards integration in Europe in forest matters justifies a European-scale approach, taking into account specific national circumstances. However, the sovereignty of each country in forestry

Fig. 2.1. The 30 European countries included in the scenario studies of the present study (dark shading).

matters remains undisputed. Neither do we challenge the fact that some countries have their own forest resource analyses methods and models, and are very well able to project their forest resource taking into account specific national expertise and circumstances (Nabuurs and Päivinen 1996).

And even if we forget the added value of harmonised scenarios for joint activities of European countries, the produced information may help the countries confirm their own strategies and assess their position compared to other countries regarding the forest management and wood production.

Here we remind the reader that the scenarios presented in this study will not try to predict what will happen. All the scenarios are of the type 'what-if', trying to describe the consequences of different management policies in combination with different demanded harvest of wood, based on what we know about forest development under certain circumstances.

3. INVENTORY DATA AND MODELLING APPROACHES

3.1. Input data

The input data to initialise the state of the forest resources in each country are based on national forest inventories. Institutes conducting forest inventory at the national level were asked to provide forest resource data in 1996. A detailed forest resource database (EFISCEN's European Forest Resource Database, EEFR) was established from all data that were sent. The database can be found at http://www.efi.fi/projects/eefr.

The slightly different way in which inventories are carried out, the type of data delivered, and the quality differences between countries, required the use of three different approaches in the rest of the study (see also Figure 7.3). These three together can be regarded as the European Forest Information Scenario Model (EFISCEN), where the first is the main used approach:

1. If data were provided by age class, the even-aged age class area matrix approach was used (see Section 3.2);
2. If data were provided by diameter class, the diameter distribution (uneven-aged) approach was used. This was the case for Spain as a whole, and parts of Italy, France and Belgium (Section 3.3);
3. If only highly aggregated data were received or available, a simple balance method was used. This was the case for Bosnia and Herzegovina and Greece (Section 3.4).

A new dataset was received from some countries, and in these cases the data that were used in the present study are the same as the ones of the IIASA Forest Study (Nilsson et al. 1992). The present version of EEFR database contains 2689 forest types in Europe (Schelhaas et al. 1999, Nabuurs et al. 2000, Pussinen et al. 2001, Nabuurs 2001) and covers 138.8 million ha of the exploitable forest area in Europe (Table 3.1).

The data in the EEFR database usually covered the exploitable forests according to UN-ECE/FAO (1992). However, there is consid-

Table 3.1. Brief overview of the EEFR database at the country level and comparison of forest area to UNECE/FAO (1992, 2000).

Country	Forest area (1000 ha) (UN-ECE/FAO 2000)	Exploitable forest area (1000 ha) (UN-ECE/FAO 1992)	No of forest types in EEFR database	Year forest inventory	Area covered in EEFR database (1000 ha)
Albania	1030	910	16	1991	898
Austria	3840	3330	32	1986–1990	2942
Belgium (Flanders)	646	620	16	1980s	110
Belgium (Walloon, even-aged)			14	1980–1984	220
Belgium (Walloon, uneven-aged)			4	1980–1984	201
Bosnia and Herzegovina	2276		8	1980s	733
Bulgaria	3590	3222	12	1980s	3202
Croatia	1775		8	?	1443
Czech Republic	2630		14	1983–1993	2584
Denmark	445	466	35	1990	442
Finland	21720	19511	64	1986–1994	19752
France (even-aged)	15156	12460	35	1983–1993	8422
France (uneven-aged)			20	1983–1993	4896
Germany	10740	9852	117	1986–1990/1993	9979
Greece	3359	2289	235	1961–1987	3252
Hungary	1811	1324	15	1995	1609
Ireland	591	394	35	1992–1993	329
Italy (even-aged)	9857	4387	19	1985	3832
Italy (uneven-aged)			30	1985	1925
Luxembourg	86	82	6	1989	71
Macedonia	906		8	?	653
Netherlands	339	331	14	1985–1992	301
Norway	8710	6638	408	1986–1993	7145
Poland	8942	8460	160	1993	8703
Portugal	3383	2346	16	1992	1613
Romania	6301	5413	34	1980s	6211
Slovak Republic	2016		84	1994	1823
Slovenia	1099		10	1991–1995	1077
Spain	13509	6506	810	1986–1995	13905
Sweden	27264	22048	150	1988–1992	22175
Switzerland	1173	1093	60	1983–1985	1044
Turkey	9954	6642	157	1990	5466
United Kingdom	2469	2207	27	1979–1982	1930
Yugoslavia, Fed state of	2894		16	?	1511
Total	**168511**	**120531**	**2689**		**138888**

erable variation, because sometimes only the state owned forests could be reported at the desired level of detail, or because special uses were excluded. Furthermore, the inventory year of the input data may vary by country, depending on the availability of data. On average the inventory data that were used for the initial year present the state of the European forests for 1988 (see also Figure 7.1), but it varies from country to country: from 1974 (average for Greece) to 1995.

3.2. The age class area matrix approach

The European Forest Information Scenario Model (EFISCEN) described in more detail in Pussinen et al. (2001) and Nabuurs (2001), is mainly an area-matrix model. The other two approaches are described in Sections 3.3 and 3.4. The core of the area matrix model is based on Sallnäs (1990) and Nilsson et al. (1992). The area matrix model is suitable for projections of forest resource of large areas under assumptions of forest management. The projections carried out with this model provide insight into increment, growing stock, age class distribution, and actual fellings for tree species and regions in a country. The EFISCEN model uses time steps of five years. The input inventory data are structured by forest types, which are defined by country, region, owner, site class, and tree species. The data of each forest type contains the following variables by age classes:

- area (ha);
- average growing stock (over bark, m^3/ha);
- increment (over bark, m^3/ha/yr).

It was not always clear whether gross annual increment or net annual increment had been received from the country correspondents. By comparing to UN-ECE/FAO (2000), most plausible was to assume that for all countries the gross annual increment was received. See section below on how that determined the way in which the mortality was simulated.

The details of the area matrix model can be found in Pussinen et al. (2001). The state of forest is depicted as an area distribution over age and volume classes in a volume-age matrix (Figure 3.1). A separate matrix is set up for each forest type of the inventory data. The projection of increment in the model is based on growth functions that are calibrated based on the inventory data. In the matrixes growth

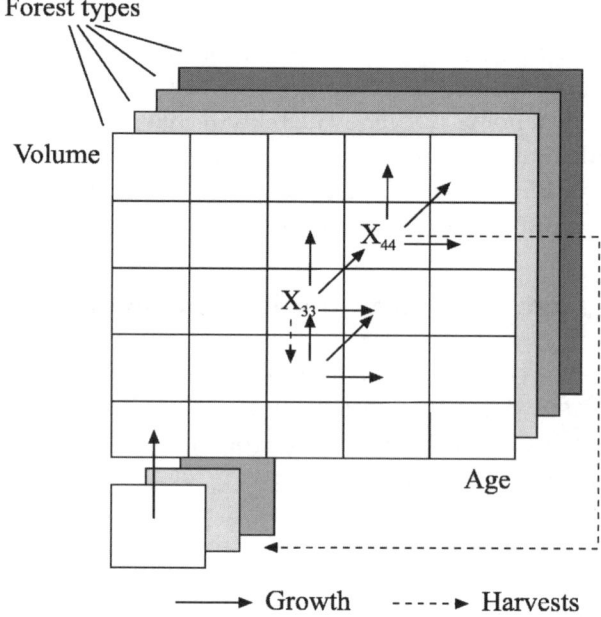

Fig. 3.1. The even-aged age class area matrix approach (Nilsson et al. 1992).

is represented as a probability of the area to grow to higher volume classes. The inventory data usually represent the situation in a country in between the mid-1980s to the early 1990s; i.e. the projections assume that the increment has not changed and will not change during the simulation period.

Ageing of the forest is incorporated as a function of time up to the point of clear cutting. Forest management is controlled at two levels in the model. First, a basic management for each forest type, like thinning and final felling regimes, is incorporated. These regimes are seen as constraints of cutting levels. The thinning regimes are incorporated as the range of age classes at which a thinning can be carried out in each forest type. Final felling regimes for each age class and forest type are incorporated as a probability that a final felling can, in principle, be carried out. Second, the required total volume of harvest was specified for the whole country for conifers, deciduous and coppice tree species groups for each time period. Thinning is carried out in the matrix of each forest type by preventing part of the area in a cell from moving to a higher volume class; i.e. growth of forest is thinned.

Natural mortality is described as a percentage of the area in a cell moving one volume class down in the matrix (Schelhaas et al. 2002). In the scenarios it was parameterised at 1% per 5 years of all areas in stands up to 150 years of age, increasing by 0.25% for every 10 years age class over the age of 150 years. Furthermore, in the top volume class of the matrix these probabilities were multiplied by a factor of 3 to describe increased mortality in dense, highly stocked stands. This parameterisation was tested and it resulted in realistic natural mortality rates of 4–15% of the gross annual increment in a country, depending especially on trends in that country's growing stock (Schelhaas et al. 2002, Harmon et al. 1986, Hees and Clerkx 1999).

3.3. Diameter distribution based approach

For parts of Belgium, Italy, France and the whole of Spain (i.e. 20.9 million ha altogether) a diameter distribution approach was used that is comparable to the method of Buongiorno and Gilless (1987). The approach used here was initially developed by Guo (Guo and Bouton unpubl.) and was applied to the forest inventory data where the forest state is described by the distribution of the number of trees over diameter classes. The model that was used can be characterised as a linear stage class model. Since this model has not been published before, a more detailed description is given here.

Based on the growth rate obtained from the inventory data, the trees move from one stage to the next with a certain transition rate independent of their location in the stand relative to their neighbours. Under scenarios of thinning and final harvesting regime, trees are removed and regenerated again based on a recruitment rate. Given the type of data that was obtained for the uneven-aged forest of Europe, a rather simple approach that was not too data intensive was required. The linear stage class model fulfils that requirement because the transition rate and recruitment rates are not dependent on the development of the forest density during the simulation. Demanding a dynamic change of the transition rate depending on the stand density would require much more insight and data on the tree inter-relationships in the stand. Here, all the trees in a diameter class are assumed to be identical.

The model divides each diameter class into two stages: (1) the first one is for trees entering the class – it is empty at the start; and (2) the second stage is for trees already in that class – it diminishes with time when trees grow and move to the next class. Cuts and mortality in a given diameter class decrease the number of trees of each of the two stages, in proportion to its content. The state of a particular diameter class at any time is described by the sum of the two stages. The transition of trees between classes works from the second stage of the lower class to the first stage of the upper class. If the number of trees at the beginning in the second stage of the lower class is No, and if the diameter increment in this lower class is I, then:

$$N = No * I/50 = No/T \qquad (1)$$

where N is the number of trees that go each year from the lower class to the upper class, and where T is the transition time in years. This number is only affected along time by cuts and mortality. The same procedure is applied during T years. The second stage of the lower class decreases each year by $N(t)$ when the first stage of the upper class increases each year by $N(t)$. After T years, the second stage has lost all the trees that have gone to the next diameter class. The contents of the first stage are then 'poured' into the second stage; a new No is given, and the process starts again. Note that the time T is different from one diameter class to another.

The number of trees in the first stage of the diameter class i ($N1_i(t)$) is a summation of the number of trees that was already in that stage the previous time step ($N1_i(t-1)$), plus those trees that have moved into that stage ($IN_i(t-1, t)$), minus those that have moved to the following stage $UT12_i(t-1, t)$, minus the number of trees being cut ($CN1_i(t-1, t)$), minus the number of trees which died due to natural mortality ($DN1_i(t-1, t)$).

$$\begin{aligned} N1_i(t) = N1_i(t-1) + IN_i(t-1, t) - UT12_i(t-1, t) \\ - CN1_i(t-1, t) - DN1_i(t-1, t) \end{aligned} \qquad (2)$$

The number of trees in the second stage of diameter class i ($N2_i(t)$), is calculated in a comparable way:

$$\begin{aligned} N2_i(t) = N2_i(t-1) + UT12_i(t-1, t) - UT_i(t-1, t) \\ - CN2_i(t-1, t) - DN2_i(t-1, t) \end{aligned} \qquad (3)$$

DIAMETER DISTRIBUTION BASED APPROACH

The aggregated number of trees in the diameter class i: ($N_i(t)$) consists simply of a summation of (2) and (3):

$$N_i(t) = N1_i(t) + N2_i(t) \tag{4}$$

Flux variables: The transition rate of the number of trees (trees/year) from the diameter class i (n = number of diameter classes) to the upper diameter class $i+1$ ($UT_i(t-1, t)$) is calculated as the number of trees in the upper stage of the diameter class (N2) divided by the transition time (TR):

$$UT_i(t-1, t) = N2_i(t-1)/TR_i(t-1),$$
$$\text{for } i = 1, 2, \ldots, n-1; \tag{5}$$
$$UT_i(t-1, t) = 0, \quad \text{for } i = n$$

where $TR_i(t-1)$ is the average delay (years) for a tree to move from diameter class i to the upper diameter class $i+1$. It is estimated by the diameter increment for the class i ($Idi(t-1)$, (cm/tree year)) in the following way:

$$TR_i(t-1) = DW/Idi(t-1) \tag{6}$$

where DW is the diameter class width, which was 5 cm for French data, and 10 cm for Italian, Spanish and Belgian data.

The entry rate of the number of trees (trees/year) into the diameter class i ($IN_i(t-1, t)$) is simply set to equal the number of trees that have left the previous diameter class:

$$IN_i(t-1, t) = UT_{i-1}(t-1, t), \quad \text{for } i = 2, 3, \ldots, n \tag{7}$$

For $i = 1$, i.e. the first diameter class, $IN_i(t-1, t)$ is determined by the recruitment rate which is derived from the inventory data.

The transition of the number of trees (trees/year) from the first stage into the second stage of the diameter class i ($UT12_i(t-1, t)$) is treated in a different way, i.e. as a pulse function where the complete number of trees moves from the first stage to the second at each time interval for the transition rate:

$$UT12_i(t-1, t) = N1_i(t-1),$$
$$\text{for each time interval of } TR_i(t-1); \tag{8}$$
$$UT12_i(t-1, t) = 0, \quad \text{otherwise.}$$

This means that there is no annual (or other time step) movement of trees from the first stage to the second, but all trees move at once at the end of the transition time, precisely at the time that the second stage has become empty.

The number of trees cut per year in each of two stages of the diameter class i ($CN1_i(t-1, t)$ and $CN2_i(t-1, t)$) is calculated according to a reference rate of number of trees cut in the diameter class i (R_i), which is an exogenous variable:

$$CN1_i(t - 1, t) = R_i * N1_i(t - 1) \qquad (9)$$

$$CN2_i(t - 1, t) = R_i * N2_i(t - 1) \qquad (10)$$

R is also derived from the inventory data. The total number of trees cut in the diameter class i ($CN_i(t)$) is simply a summation of (9) and (10):

$$CN_i(t) = CN1_i(t) + CN2_i(t) \qquad (11)$$

The number of trees that die through natural mortality (suppression, etc.) per year in each of the two stages of the diameter class i ($DN1_i(t-1, t)$ and $DN2_i(t-1, t)$) is incorporated as a percentage of the number of trees (D_i) in each stage D_i and is treated as an exogenous variable:

$$DN1_i(t - 1, t) = D_i * N1_i(t - 1) \qquad (12)$$

$$DN2_i(t - 1, t) = D_i * N2_i(t - 1) \qquad (13)$$

The total number of trees that died in the diameter class i ($DN_i(t)$) is simply a summation of (12) and (13):

$$DN_i(t) = DN1_i(t) + DN2_i(t) \qquad (14)$$

The recruitment rate is the number of trees that enter the first class each year. It can be estimated from the number of trees in the first diameter class and the transition time for this class:

$$RCM = NI(D10)/TR(D10) + SUM(NI) - N5 \qquad (15)$$

where the latter part of the equation ($SUM(NI) - N5$) is a correction factor that takes into account the change in the number of trees compared with the initial situation. If the number of trees increases $N5 > NI$, then the summation is negative and recruitment will be suppressed. The translation from number of trees to the total volume is simply a multiplication with the average tree volume, which comes from the inventory data.

3.4. Simple balance approach

The data available for Bosnia and Herzegovina and Greece were not suited to either of the approaches described above. Therefore, a simple balance method was used as was done by Kuusela (1994) and later by Päivinen et al. (1999b) as a verification method for the Leningrad Region. In this method the growing stock (average for the whole country) in the following time step is simply calculated as the initial growing stock, plus gross increment, and minus fellings and natural mortality. Increment was parameterised at 2.5% if the growing stock was less than 150 m^3/ha, and at 2% when the growing stock was higher. Natural mortality was assumed to be 8% of increment when the growing stock was less than 150 m^3/ha, and 10% of increment when the growing stock was higher.

4. SCENARIOS

4.1. Basic assumptions made

EFISCEN does not predict future total fellings. It only distributes total demand in terms of fellings overbark (o.b.) over a dynamically growing forest resource. The total demand development over time may be based on information that comes from market model studies. The fellings are distributed over the forest types, age and volume cells according to probabilities specified for each cell. It is assumed that this provides information on the state of the forest resource at a sufficiently detailed level.

In deriving the scenarios, the following basic assumptions have been made:

Growth

- It was assumed that the forest inventory data as supplied by countries represent the state of the national forest according to the forest area as given by FRA1990 (UN-ECE/FAO 1992) or (UN-ECE/FAO 2000) in sufficient detail to model growth, age development, etc. of the forest. Furthermore, it is assumed that forest growth can be sufficiently accurately modelled at the minimum area of 10 ha for each cell in each matrix.
- Environmental changes, such as climate change, pollution, CO_2 fertilisation and N deposition, will not further change the site fertility. Thus, growth rates will stay at the level of the late 1980s, because the growth functions were fitted on data representing that period. It is assumed that these functions will stay valid until 2050.
- It was assumed that the tree species distribution of the forests will not change over the time span of the simulated 60 years[1].

---

[1] Towards the end of this study, a possibility to change tree species distribution was developed for EFISCEN. However, that was not applied in the current study. It seems reasonable to assume that tree species changes will hardly have an impact on harvesting potential per species (group) within the simulated 60 years.

Mortality

- In relation to environmental changes, it was assumed that large-scale forest decline does not occur. Natural mortality will remain a function of age and growing stock volume. Since the input data represented gross annual increment, any natural mortality occurring will be at the cost of the 1990s gross annual increment.
- It was assumed that there will be no change in disturbance rate, and practically all occurring natural disturbances will be salvaged and taken up in normal planned harvesting as has been the case in the past.

Demand

- It was assumed that no major changes will occur in terms of development of new technologies to substitute the use of wood-based products, neither will major economic catastrophes occur that may hamper the utilisation of wood resources as anticipated.
- In relation to this, we assume that forestry stays financially attractive and owners will continue to be willing to carry out (some) harvesting activities. They will show the same price elastic behaviour as in the past. This also means that there will be no major shift of harvesting to Russia.
- It was assumed that there will be no major forest area expansion in Europe

Management

- It was assumed that there will hardly be any technological improvement in felling methods, and its associated costs. The regeneration will take place with the same species as the one that was clearcut, and thus a species distribution change will not occur. The success and methods of forest regeneration stay the same.
- For most scenarios (business-as-usual, maximum sustainable production and ETTS) it was assumed that the management regimes of final felling and thinning per owner and tree species will be the same as in the 1990s.
- It was assumed that there would be no shift in harvesting intensity from one owner group to another.

4.2. Scenarios

Based on the issues outlined in Chapter 1, the following four scenarios were drawn up – business as usual, EFISCEN ETTS-V, maximum sustainable production and multi-functional. Scenarios were run from the period represented by the inventory data until 2050.

In the business-as-usual scenario (BAU) we assume that the present (1990) total national felling levels will stay the same throughout the simulation period. Felling levels are specified by coniferous and deciduous species groups and by thinning or final cut. Management regimes are applied as they are today. Both the species distribution and the total forest area do not change. This scenario addresses the question of what will happen to the European forests if management continues as it is today.

In the EFISCEN ETTS-V (EFISCEN-ETTS) scenario we use the felling levels as they were projected by the UN-ECE/FAO European Timber Trend Studies V for the period 1990–2040 (Pajuoja 1995). The aim is to assess the forest development (e.g. growing stock, age classes and increment) under the felling levels as the UN-ECE foresaw them. It addresses the question of the accuracy of these kind of large-scale projections, and differences that may occur when using another methodology. Felling levels are specified by coniferous and deciduous species groups and by thinning or final cut. Management regimes are applied as they are today. The species distribution does not change during the run, but forest area expansion is incorporated as was foreseen by the ETTS-V study.

In the maximum sustainable production (MAX) scenario the search is for a maximum felling level under which the ratio between increment and fellings is close to one, and average standing volumes are thus maintained at present levels[2]. To search for that level, usually the total required felling was increased quickly. Both the species distribution and the total forest area do not change. This scenario addresses the question of what would be the biological theoretical maximum sustainable felling level under which the forest does not degrade, if the European forests would have to take up the increased pressure on the forest resource as described in Chapter 1.

---

[2] A very simple interpretation of 'sustainable' was thus adopted here.

In the multi-functional (MTF) scenario we assume that the current trend in forest management towards a more nature-oriented management (see Chapter 1) continues in the future. With this scenario we assess whether the usually foreseen rise in felling levels of around 0.5–1% per year can be sustained in the future. Generally the foreseen fellings are increased in these scenario runs with the mentioned percentage and more attention is paid to the nature values of the forests in the country. The foreseen rise in fellings is a combination of:

1) an expected decline in fellings due to the fact that owners will not put so much emphasis on wood production in the future; and
2) the ongoing discussion concerning bio-energy potentials of European forests and the aim of increasing the proportion of bio-energy coming from (pre-commercial) thinnings, logging slash, etc. Based on the combination of these two developments we assume that fellings will increase by 0.3% per year for the first 30 years, and then stabilise.

The incorporation of nature-oriented management is done through:

A) setting aside the older age classes from final felling as of 1990 (it depends on the country which age classes and in which forest types). This means that all the forest area in those age classes, and all areas moving into that age class during simulation, cannot be clear cut anymore. This we call the area of set-aside (note that these set-aside areas are scattered over the country, and do not have a specific location);
B) additionally it is assumed that a larger proportion of the total fellings has to originate from thinnings (usually up to 45% by 2050). This is to represent a move towards more selective cutting systems instead of clearfellings;
C) then it is also specified that the average rotation length in all forest types is prolonged by one age class (usually 20 years), and the age at which thinnings can be carried out is also prolonged by one age class;
D) as a result of marginal agricultural land becoming available, gradual forest area expansion of, in total, 4.4 million ha is incorporated in this scenario.

5. COUNTRY LEVEL RESULTS

5.1. ALBANIA

In co-operation with Prof. A. Postoli

Introduction

Albania is rich in forests; 37% of the land is under forest and other wooded land, and just over half of this is coppice and shrubland. Due to over-exploitation, fire and grazing the forest area has been reduced by 20% over the last 40 years (Meta 1993). Most of the 1 million ha of forest and other wooded land consists of broadleaved species, giving the country a natural species composition. Most of the forest can be characterised as mountainous forests with a low net annual increment and a low average growing stock. This seems to deviate strongly from the potential, given the favourable growing conditions. All of the forests are under public ownership.

Country specific scenario assumptions

The applied input data are based on UN-ECE/FAO (1994), and it was assumed that those data represent the situation around 1990. A more detailed, or more up-to-date dataset was not obtained from the country directly. Plans for a new forest inventory are underway at present. The input data distinguished 16 tree species of which six are of a coppice type structure. These data cover 898 000 ha of forest. UN-ECE/FAO (1992) gave 910 000 ha as exploitable forest, thus the input data covered 98% of the latter area.

Business-as-usual scenario

The projections made here were mostly based on official statistics. UN-ECE/FAO (1994) gave a total annual felling level of 0.41 million m^3 overbark (o.b.), of which 35 000 m^3 is from conifers. Still, the current fellings level in Albania is rather uncertain, as official statistics may underestimate fuelwood collection, which is not necessarily recorded. This level was used as the assumed fellings level throughout the simulation period. The COMTRADE removals data also show a rapid decline to 0.409 million m^3. UN-ECE/FAO (2000) estimate a total annual felling level of 0.74 million m^3, of which 80 000 m^3 is from conifers.

Two-thirds of the deciduous forest is managed as coppice, and hence the same proportion was assumed for the fellings from coppice: required fellings in deciduous high forest are 0.12 million m^3/yr, and fellings from coppice 0.26 million m^3/yr. The required proportion of thinnings from total fellings was assumed to be 28%. It was assumed that the area of forest available for wood supply will stay at the 1990s level.

EFISCEN-ETTS scenario

ETTS-V reported much higher fellings (2.1 million m^3/yr) than the BAU scenario. This annual felling was divided between the forest type groups in the same way as in the previous scenario.

Maximum sustainable production scenario

In 10 years time, the fellings were increased gradually to a level under which the growing stock remains approximately stable.

Multi-functional scenario

In this scenario it was assumed that the forest sector of Albania will recover from the current economic recession. Hence fellings start to increase rapidly in deciduous high forest and coppice after 5 years, and continue to increase until 2030 to a level that is 75% of the ETTS-V levels. After that fellings stabilise. It was assumed that the past degradation and reduction in forest land will be curbed, and thus the current forest area remains stable.

Management regimes were adapted in order to pay more attention to biodiversity values. The species distribution was kept as it was in 1990, but the following forests were set aside for reserves: high forest of oak, beech and chestnut older than 140 years and shrub woodlands older than 60 years. The rotation length was increased by 20 years for all forests. Also, the proportion of thinnings from total fellings is gradually increased in deciduous high forest to 50%.

Results for Albania

The high degree of uncertainty about actual fellings at present in Albania makes the projections rather uncertain. A fast decline in fellings in the 1990s has occurred according to official statistics. Starting from that low present level and continuing that in the BAU scenario leads to a build-up of growing stock to 125 m^3/ha.

ALBANIA

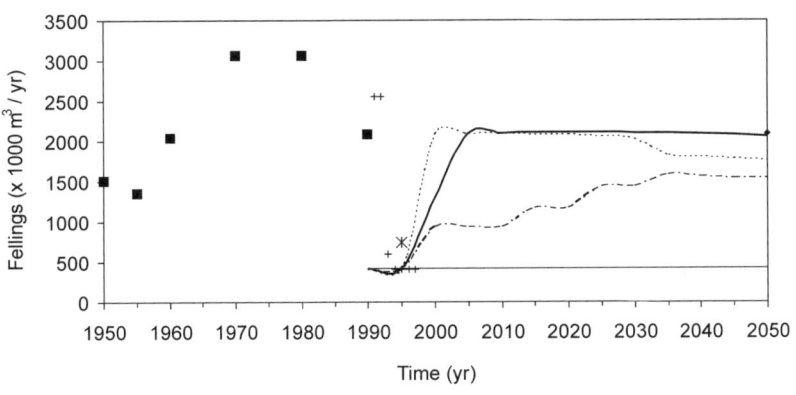

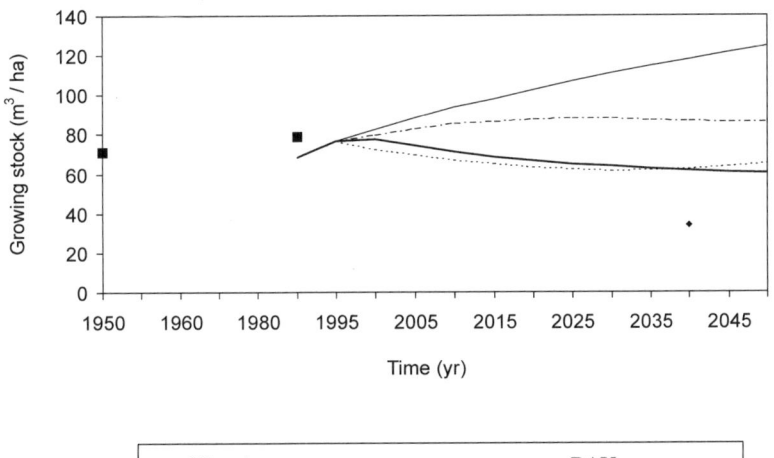

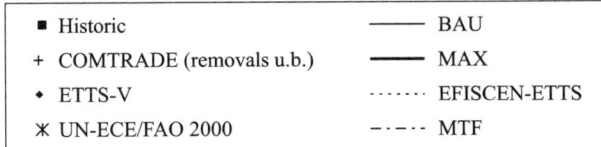

Fig. 5.1. Projected total national fellings (top) and growing stock per ha (bottom) in Albanian forests under the four scenarios until 2050.

Given the simulated gross annual increment of around 2.3 m³/ha/yr, a maximum sustainable production of 2.1 million m³/yr may be achieved. Under this felling level, the average growing stock remains around the level of 67 m³/ha. The stocking developing under this scenario leads to slightly declining natural mortality rates from 11% of the gross annual increment in 1990 to 7% in 2050.

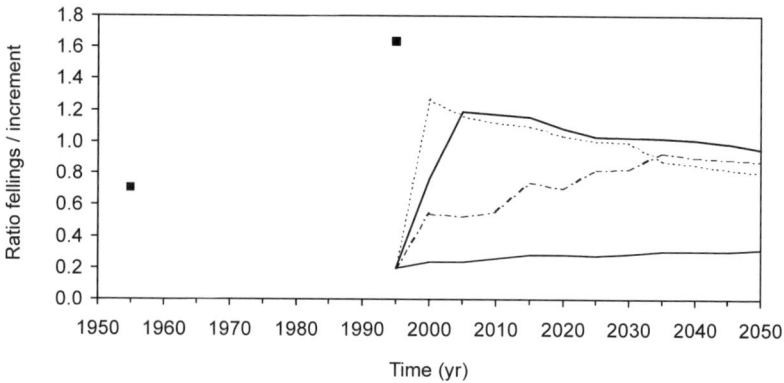

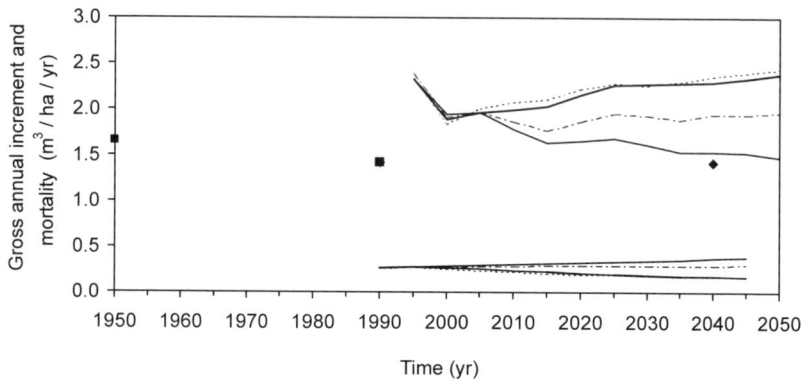

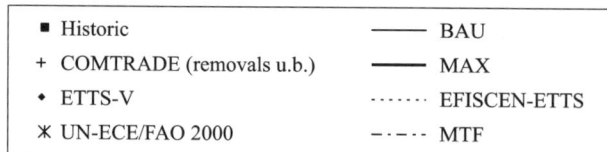

Fig. 5.2. Ratio of projected fellings/increment in Albanian forests under the four scenarios until 2050 (top) and projected net annual increment and mortality (bottom).

The ETTS V study had foreseen that fellings in Albania would remain at the level of the 1980s, around 2 million m^3/yr. Despite the higher gross annual increment in the present study, it was not possible to maintain that level in 2030–2050. After 2030 actual fellings start to decline in EFISCEN-ETTS, and throughout the whole simulation

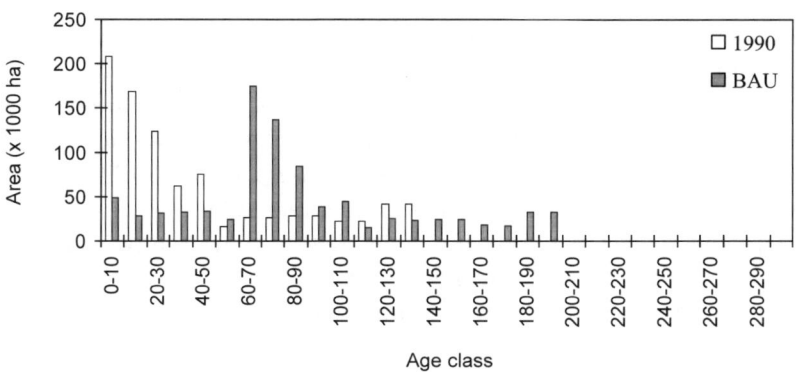

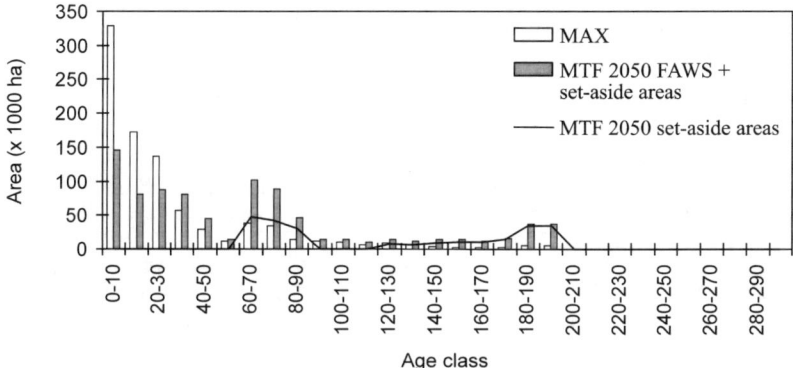

Fig. 5.3. The age class distribution of the Forest Available for Wood Supply (FAWS) in the initial situation and for 2050 under the BAU scenario (top). The age class distribution of FAWS for 2050 under the MAX scenario, the MTF scenario including set-aside areas and the MTF scenario with only set-aside areas (bottom).

period a decline in growing stock occurs to around 61 m^3/ha by 2030. The main difference between the MAX and the EFISCEN-ETTS scenarios is the higher fellings in the latter in the late 1990s, causing lower fellings in 2030–2050.

Albanian forests can meet the projected fellings under the multifunctional scenario in which it was assumed that the economy will recover, and that felling will gradually increase to 1.5 million m^3/yr by 2035. At that time a felling/increment ratio of 93% is reached, which seems to be the maximum. Mortality rates may then amount to 15% and the growing stock has increased to 88 m^3/ha.

Under the three scenarios in which the highest fellings are reached, the gross annual increment remains approximately at the level of the 1990s: 2 to 2.4 m^3/ha/yr. The BAU scenario leads to older forests, and thus a greater decline in gross annual increment.

The age class distribution for 1990 clearly shows the proportion of coppice. Under the BAU scenario, much of that coppice area may have outgrown to high forest, as indicated by the area in the classes 60–90 years.

The MTF scenario shows that under the assumed felling levels a significant area of older forests has developed, of which 249 000 ha are projected to be newly established as set-aside areas in the MTF scenario. There were no problems in finding the required fellings.

REFERENCES

Meta, M. 1993. Forests and Forest Policy in Albania. Journal of Forestry 91(6): 27–28.

UN-ECE/FAO 1992. The Forest Resources of the Temperate Zones, the UN-ECE/FAO 1990 Forest Resource Assessment, General Forest Information. Geneva, Switzerland. 347 p.

UNECE/FAO 1994. Forest and Forest Products Country Profile: Albania.

UN-ECE/FAO 2000. Forest Resources of Europe, CIS, North America, Australia, Japan and New Zealand. Geneva, Timber and Forest Study Papers, No. 17. 445 p.

5.2. AUSTRIA

In co-operation with D.I. Karl Schieler

Introduction

Austria is a land of forests; it is the third most densely forested country in the EU. Of the 3.84 million ha of forest (mostly privately owned), 62% consist of Norway spruce. Most of the forest can be characterised as mountainous forests with a high net annual increment and large average growing stock. Despite the large share of land being covered by forests (47%), its forest sector (forestry and industry) contributes only 3.8% to the national GDP (Czamutzian 1999).

Country specific scenario assumptions

The forest inventory data that were used as input to the simulations are based on the 1988 inventory data (Schieler and Büchsenmeister 1995). These data covered 2.94 million ha that are exploitable high forest, and excludes the 'Protection forest under active management'. Therefore, the simulations have covered 89% of what is recorded in Austria as forest available for wood supply. The input data were divided into eight regions, four owner groups, one site class and one tree species.

Business-as-usual scenario

Several statistics were consulted to set a required fellings level. The fellings over bark provided by UN-ECE/FAO (1992) were 17.4 million m^3 year. The COMTRADE data on removals underbark (u.b.) data for 1990–1997 fluctuate between 16.7 and 12.8 million m^3/yr, while UN-ECE/FAO (2000) reports annual fellings overbark (o.b.) of 19.5 million m^3/yr. Since the input data cover 89% of the exploitable forest area, in this scenario a total felling of 17.4 million m^3/yr was assumed throughout the simulation period. Out of this amount, 15% comes from thinnings.

EFISCEN-ETTS scenario

ETTS-V foresaw an increase in fellings of 25% in the period 1990–2030. This trend was followed here, starting with the fellings as used in the BAU scenario. Out of this amount, 15% comes from thinnings.

ETTS-V foresaw an expansion of the forest area. This was followed in this scenario with a forest area expansion of 64 000 ha over a period of 40 years.

Maximum sustainable production scenario

The fellings were increased in 15 years time to a level under which the growing stock remains approximately stable.

Multi-functional scenario

The fellings are foreseen to increase by 1% per 5 years, starting with the fellings as used in the BAU scenario. The increase will continue until 2030. The management regimes were adapted in order to pay more attention to current trends in forest management. All forests older than 150 years were taken out of production. Also the rotation length of all species is prolonged by 20 years and the share of thinnings in total fellings is gradually increased to 50%. The species distribution was kept as it was in 1990. The forest area was expanded by 64 000 ha over a period of 40 years.

Results for Austria

The projections for 2.94 million ha of forests show that a maximum sustainable production of 23.1 million m^3/yr may be achieved. Under this fellings level, average growing stock declines slightly towards the end of the simulation period, which seems tolerable given the present large growing stock. The stable stocking developing under this scenario also leads to stable natural mortality rates of some 7.5% of the gross annual increment.

The Austrian forests can easily meet the projected fellings under the EFISCEN-ETTS and the MTF scenarios with an annual felling of 21 million m^3/yr and 18.6 million m^3/yr, respectively, by 2050. These are felling/increment ratios of 84% and 63%, respectively. Mortality rates may then amount to 11.3% and 12.7%, respectively. The growing stocks will still be increasing. By 2050 growing stocks under the EFISCEN-ETTS and the MTF scenarios amount to 387 and 445 m^3/ha, respectively.

If fellings stay at the level of the 1990s (BAU), growing stocks may increase to 429 m^3/ha. So, despite the lowest felling level in this scenario, it does not lead to the highest growing stock, mainly due to the high growing stock of set-aside forest within the MTF scenario.

AUSTRIA

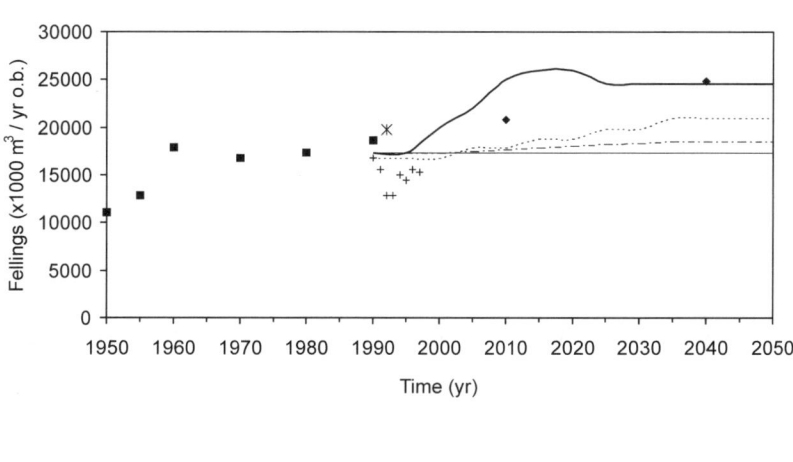

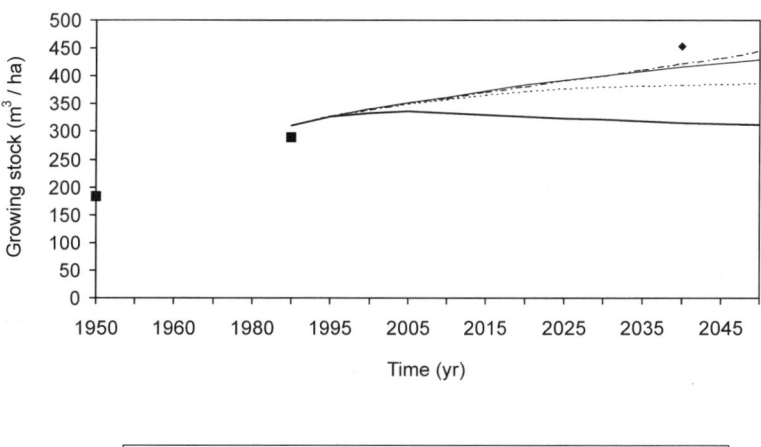

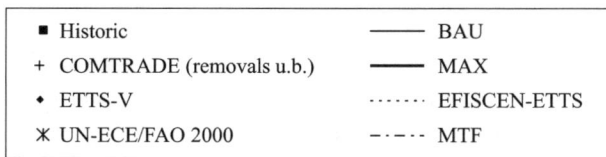

Fig. 5.4. Projected total national fellings (top) and growing stock per ha (bottom) in Austrian forests under the four scenarios until 2050.

Under all scenarios gross annual increment declines slightly due to increase in age and stocking in the forest. The decline is usually around 1.5 m^3/ha/yr over the simulated period of 60 years, except for the MTF scenario where the decline is only 0.5 m^3/ha/yr. This is because in the MTF scenario the projected larger share of fellings from thinnings leads to additional growth. The MTF scenario shows that

38   COUNTRY LEVEL RESULTS

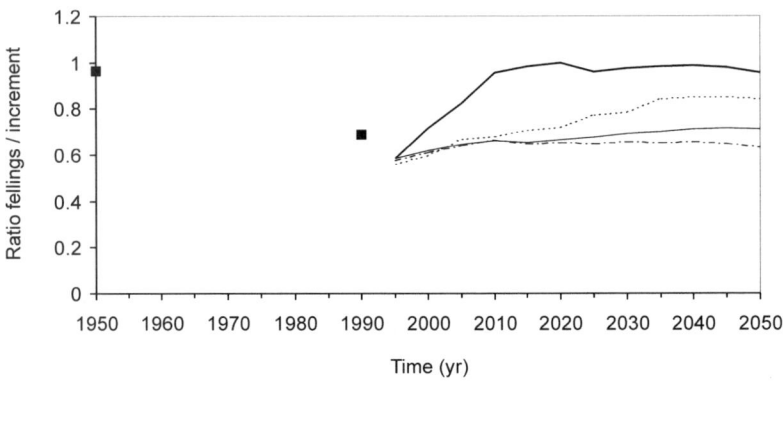

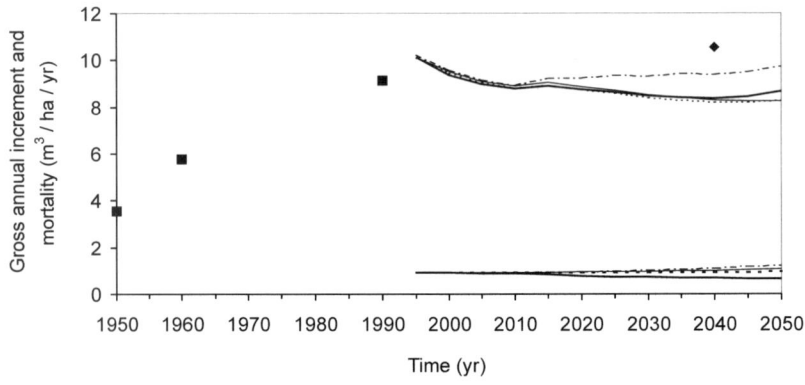

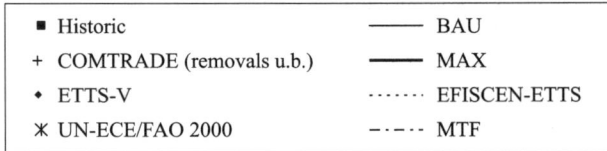

Fig. 5.5. Ratio of projected fellings/increment in Austrian forests under the four scenarios until 2050 (top) and projected net annual increment and mortality (bottom).

under the assumed felling levels there will not be a major shift in age class distribution in Austrian forests by 2050. Only a significant area of forest older than 170 years of age has developed. However, this ageing process has occurred to a comparable degree also in the BAU scenario. Out of the area older than 170 years, 331 000 ha are projected to be

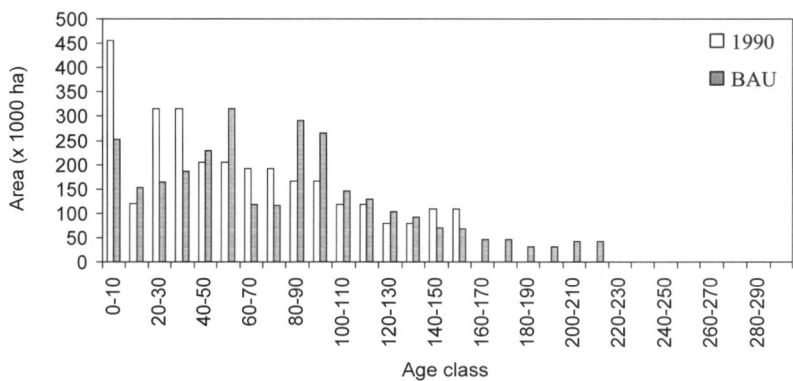

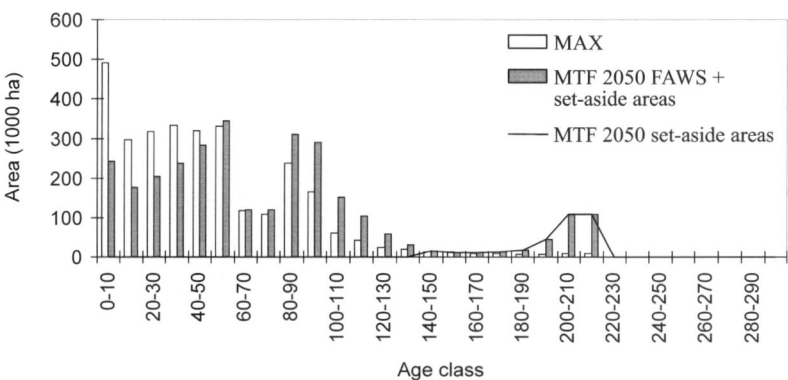

Fig. 5.6. The age class distribution of the Forest Available for Wood Supply (FAWS) in the initial situation and for 2050 under the BAU scenario (top). The age class distribution of FAWS for 2050 under the MAX scenario, the MTF scenario including set-aside areas and the MTF scenario with only set-aside areas (bottom).

newly established as set-aside areas in the MTF scenario. This does not lead to problems in finding the required fellings.

REFERENCES

Czamutzian, S. 1999. Austria. In: Pelkonen, P., Pitkänen, A., Schmidt, P., Oesten, G., Piussi, P. and Rojas, E. (eds.), Forestry in Changing Societies in Europe. Part II. SILVA Network. University Press, Joensuu, Finland. Pp. 1–22.
Schieler, K. and Büchsenmeister, K.S.R. 1995. Österreichische Forstinventur. Ergebnisse 1986/90. Forstliche Bundesversuchsanstalt – Waldforschungszentrum, Wien, FBVA Berichte Nr. 92. 262 p.

UN-ECE/FAO 1992. The Forest Resources of the Temperate Zones, the UN-ECE/FAO 1990 Forest Resource Assessment, General Forest Information. Geneva, Switzerland. 347 p.

UN-ECE/FAO 2000. Forest Resources of Europe, CIS, North America, Australia, Japan and New Zealand. Geneva, Timber and Forest Study Papers, No 17. 445 p.

Further information

Schwarzbauer, P. 1994. Conditional long-term forecast for Austrian solid wood product consumption and production based on a simulation model and on economic equations. Centrallblatt für das Gesamte Forstwesen 111(2): 129–142.

Flach, M. 1996. Timber supply analyses in Austria. In: Päivinen, R., Roihuvuo, L. and Siitonen, M. (eds.), Large scale forestry scenario models: experiences and requirements. EFI Proceedings 5. European Forest Institute. Pp. 39–48.

5.3. BELGIUM

In co-operation with Prof. Jacques Rondeux and Ms. Martine Waterinkx

Introduction

Belgium is one of the smaller forestry countries in Europe with 645 000 ha of forest (UN-ECE/FAO 2000). It is a country with two distinct regions: Flanders is an industrialised region, sparsely forested, with most of its forests scattered in an agricultural landscape; and Wallony is a hilly, densely forested region. Norway spruce is by far the most important forest tree species (Luyssaert et al. 1999). Especially in Wallony, the forest industry plays a role in employment in the rural areas.

Country specific scenario assumptions

For Belgium the input data covered 0.53 million ha of forest, of which 0.33 million ha is represented in age class distributions and 0.2 million ha is represented as diameter class distributions. The forest inventory data are based on the 1985 inventory for Wallony, and for Flanders on the state of the forest in the early 1980s. In 2001, a new inventory became available for Flanders, but was not used here. The input data distinguished two regions, two owner classes and eight tree species for the age class distribution data, and for the diameter distribution data: one region, two owner classes and four tree species.

Business-as-usual scenario

To assume a present felling level that could be continued throughout this scenario several statistics were available. UN-ECE/FAO (1992) gave a total annual felling level of 3.3 million m^3/yr from the total exploitable forest. The COMTRADE removal u.b. data present a sharp decrease from 5.0 million m^3/yr in 1990 (storm year) to 3.6 million m^3/yr in 1997. UN-ECE/FAO (2000) reports annual fellings in Belgium of 4.4 million m^3/yr, of which 3.1 million m^3/yr is from coniferous forest. This latter figure was used as the basis for the fellings levels. Taking into account the fact that only 85% of the total exploitable forest was covered, a fellings level of 2.4 million m^3 was assumed throughout the simulation period, out of which 27% had to come from deciduous species. The required proportion of thinnings from total fellings was assumed to be 30%. It was assumed that the area of forest available for wood supply will stay at the 1990s level.

EFISCEN-ETTS scenario

In the case of Belgium ETTS-V predicted a stable fellings level of 3.3 million m^3/yr by until 2040, of which 2.2 million m^3 is from coniferous forests. An attempt was made to reproduce that scenario (taking into account the fact that only 85% of the total exploitable forest was covered), although it could not be followed throughout the simulation period.

Maximum sustainable production scenario

The fellings were increased in 10 years time to a level under which the growing stock remained approximately stable; fellings increased quickly, but then again decreased steadily, to take into account the decreasing increment. Towards the end of the simulation this meant that annual fellings amounted to 3.2 million m^3.

Multi-functional scenario

In this scenario the UN-ECE/FAO (2000) fellings level was used as a starting point for 1990. After that it was assumed that fellings will increase by 1% per year until 2005 and then stabilise. The management regimes were adapted in order to pay more attention to the current trend in forest management towards a more nature-oriented management. We keep the species distribution as it was in 1990 data, but set aside deciduous high forests of diameter classes more than 70 cm (for diameter distribution approach) and forests of 'hardwoods' of more than 100 years (for age class distribution approach), initially representing 17 000 ha. The required proportion of thinnings from total fellings was assumed to increase gradually to 50% in 2040.

Results for Belgium

The projections for 0.53 million ha of forests show that a long-term average maximum sustainable production of 3.2 million m^3/yr may be achieved. However, a felling level of 4.5 million m^3/yr was possible during the period that gross annual increment is still rather high. Later on in the simulation, the maximum sustainable production level

BELGIUM

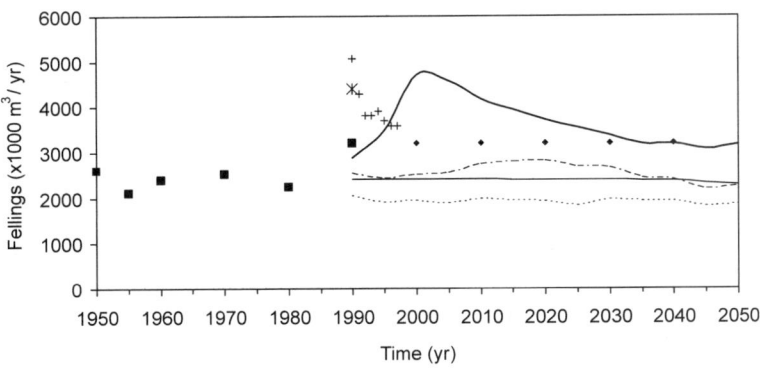

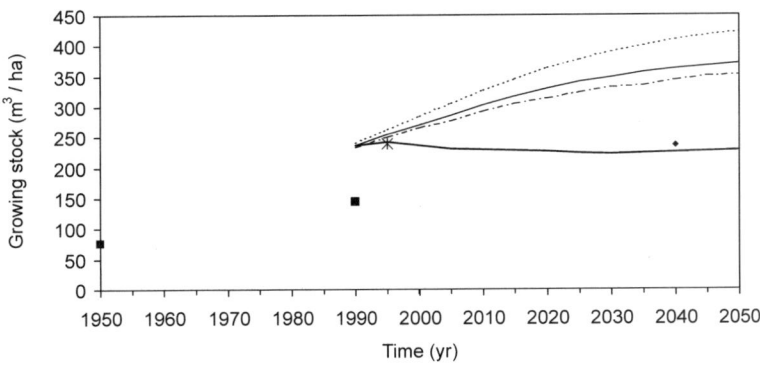

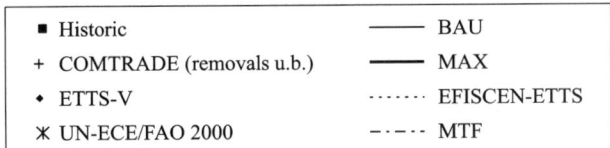

Fig. 5.7. Projected total national fellings (top) and growing stock per ha (bottom) in Belgian forests under the four scenarios until 2050.

is clearly affected by the declining increment. The stable stocking (225 m³/ha) developing under this scenario, also leads to stable natural mortality rates of some 6% of the gross annual increment.

The Belgian forests could not meet the projected fellings under the EFISCEN-ETTS scenario with a required annual felling of 2.8 million m³/yr on this 0.53 million ha. This seems strange, given

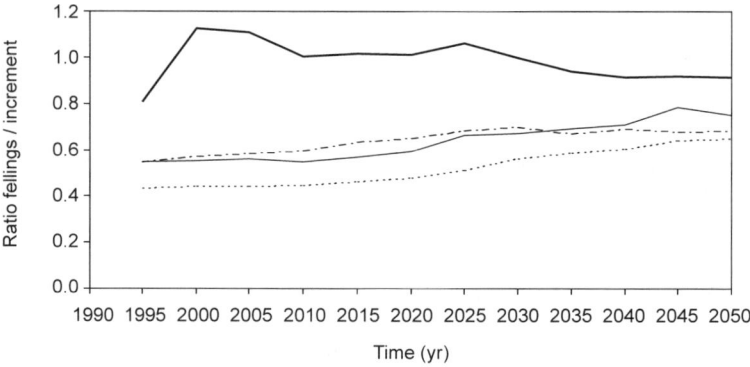

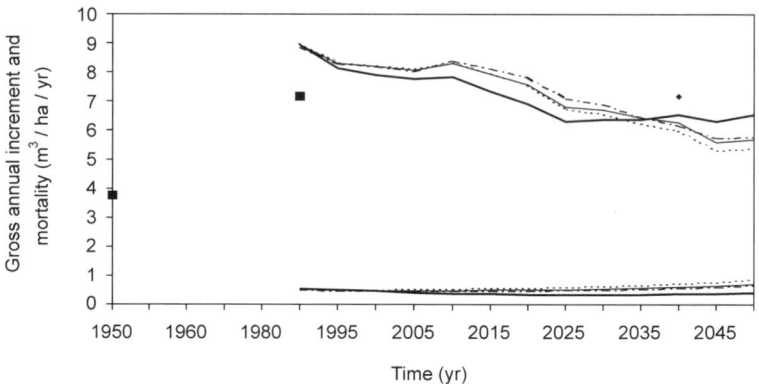

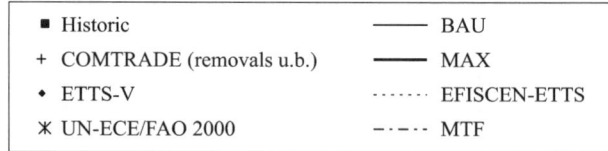

Fig. 5.8. Ratio of projected fellings/increment in Belgian forests under the four scenarios until 2050 (top) and projected net annual increment and mortality (bottom).

the higher maximum sustainable fellings level. However, in the MAX scenario, the fellings start at a much higher level, leading to sufficient regeneration and slightly higher increment (6.5 m^3/ha/yr MAX versus 5.4 m^3/ha/yr under EFISCEN-ETTS). This makes it possible to maintain a higher fellings level.

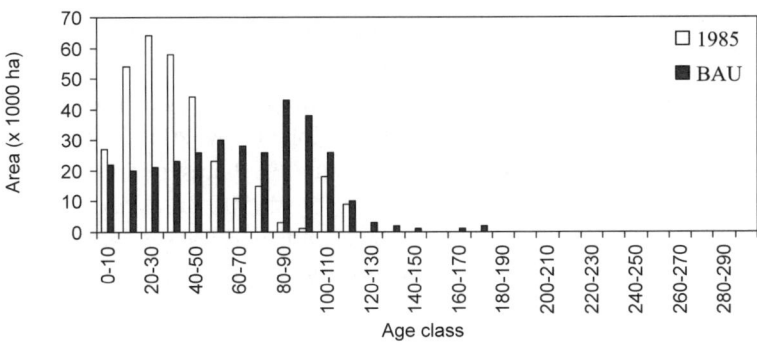

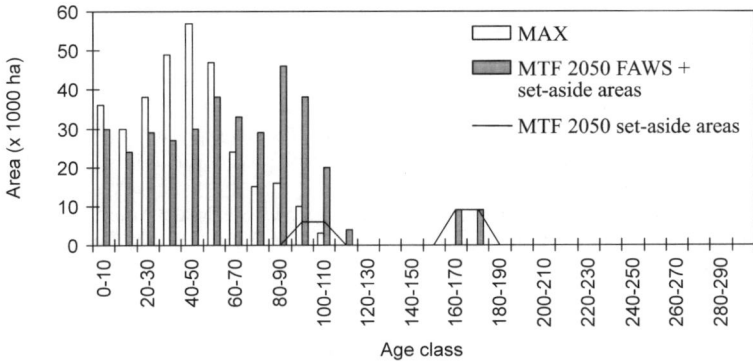

Fig. 5.9. The age class distribution of the Forest Available for Wood Supply (FAWS) in the initial situation and for 2050 under the BAU scenario (top). The age class distribution of FAWS for 2050 under the MAX scenario, the MTF scenario including set-aside areas and the MTF scenario with only set-aside areas (bottom).

It is apparent that under all scenarios the gross annual increment declines considerably due to increase in age and stocking in the forest. The decline is sharpest in EFISCEN-ETTS scenario with 3.6 m^3/ha/yr over the simulated period of 60 years, and smallest in the MAX with a decline of 2.4 m^3/ha/yr. This leads to increases in all scenarios of the felling/increment ratios from 55% in 1990 to around 70% in 2050.

If fellings stay at the level of the 1990s (BAU), growing stocks may increase to 371 m^3/ha. Mortality rates may then amount to 12% of net increment. By 2050 growing stocks under the Efiscen-ETTS and the MTF scenarios amount to 423 and 352 m^3/ha, respectively. However, in the MTF scenario, not all required fellings were found after 2020: by 2050, 0.56 million m^3/yr were not found.

Under all scenarios (except MAX) a major shift will have occurred in the age class distribution in Belgian forests by 2050. The peak of forest area in the classes 20–40 years in 1990 has decreased, but is still visible in the classes 80–100 years by 2050. By 2050 an area of 30 000 ha is projected to be newly established as set-aside areas in the MTF scenario, partially contributing to problems in finding the required fellings after 2020.

REFERENCES

Luyssaert, S., Thierron, V., André, P. and Lust, N. 1999. Belgium. In: Pelkonen, P., Pitkänen, A., Schmidt, P., Oesten, G., Piussi, P. and Rojas, E. (eds.), Forestry in Changing Societies in Europe. Part II. SILVA Network. University Press, Joensuu, Finland. Pp. 29–44.

UN-ECE/FAO 1992. The Forest Resources of the Temperate Zones, the UN-ECE/FAO 1990 Forest Resource Assessment, General Forest Information. Geneva, Switzerland. 347 p.

UN-ECE/FAO 2000. Forest Resources of Europe, CIS, North America, Australia, Japan and New Zealand. Geneva, Timber and Forest Study Papers, No 17. 445 p.

Further information

Afd Bos en Groen 2001. De bosinventarisatie van het Vlaamse gewest. Resultaten van de eerste inventarisatie 1997–2000. Ministerie van de Vlaamse Gemeenschap. 480 p.

5.4. BOSNIA AND HERZEGOVINA

Introduction

The recent changes in the political situation in Bosnia and Herzegovina explain the uncertain situation concerning the present state of the forests and its management. The country is densely forested. There is a rich deciduous forest in the mountainous regions of the Dinaric Alps (which has a Mediterranean climate). The 2.28 million ha of forest as reported by UN-ECE/FAO (2000) are mostly publicly owned and only 1.3 million ha is reported as available for wood supply. Even though UN-ECE/FAO reports no area change, it is unknown to what extent unrecorded fellings, grazing, and other forms of degradation have occurred during the 1990s. At the moment, the forest sector is in a recession in the country.

Country specific scenario assumptions

Since only a limited dataset was available from the Nilsson et al. (1992) study, it was decided to apply a simple balance method to the UN-ECE/FAO (2000) data (see Chapter 3 of this report, Pajuoja (1995), Nilsson et al. (1992) and Kuusela in Päivinen et al. (1999)). Due to unrecorded fellings occurring, it was chosen to initialise the calculation with a forest area of 2.28 million ha. An initial growing stock of 95.9 m^3/ha was used. The balance method is based on simple calculations of a gross annual increment of 2.5% of the growing stock when the growing stock is less than 150 m^3/ha. If the growing stock is higher, then a relative growth of 1.5% is used. Natural mortality is calculated in a comparable way: if the growing stock is less than 150 m^3/ha, then 6% of gross annual increment is assumed to be lost, otherwise the natural mortality is 9% of gross annual increment.

Business-as-usual scenario

There were only a few data sources that could be consulted from which a basic fellings level could be estimated. ETTS-V foresaw that fellings would increase from 60% of the increment in 1990 to 71% of the increment in 2040 (Pajuoja 1995). The 60% would mean for the present area of Bosnia and Herzegovina some 3.2 million m^3/yr. However, UN-ECE/FAO (1992) gave an annual fellings level of only 1.2 million m^3 for 1995. Since this is the most recent source of information, this amount was used and doubled as the basis for the fellings in the initial year. The doubling was assumed to account for illegal fellings. In the BAU scenario it was assumed that the fellings will stay at this level throughout the simulation period. It was assumed that the area of forest will stay at the 1990s level.

EFISCEN-ETTS scenario

In the case of Bosnia and Herzegovina ETTS-V foresaw that fellings would increase from 60% of the increment in 1990 to 71% of the increment in 2040 (Pajuoja 1995). This ratio development was followed in this scenario (i.e. with our calculated higher increments, the total fellings are also higher).

Maximum sustainable production scenario

The fellings were increased from the initial amount of 2.4 million m^3/yr to 5.8 million m^3/yr by 2015; the level under which the growing stock remained approximately stable.

Multi-functional scenario

In this scenario the UN-ECE/FAO (2000) fellings level was used as a starting point for 1990, but doubled to account for illegal fellings, i.e. 2.4 million m^3/yr. Then it was assumed that fellings will decrease to 1.9 million m^3/yr by 2010 to account for the economic recession. After that they increase by about 1% per year, and then 2% and 3% per year, assuming an economic recovery. By 2050 fellings amount to 4.3 million m^3/yr. It was assumed that the forest area continues to decrease until 2010 by 3000 ha/yr. After that an area expansion of 4000 ha/yr was assumed.

Results for Bosnia and Herzegovina

Because of the simple approach used for Bosnia and Herzegovina, all results must be regarded with care. These simple projections for 2.28 million ha of forest indicate that a maximum sustainable production of 5.8 million m^3/yr may be achieved, far higher than at present. However, the rather simple approach used here (with relative growth rate compared to the growing stock) determines the outcome to a

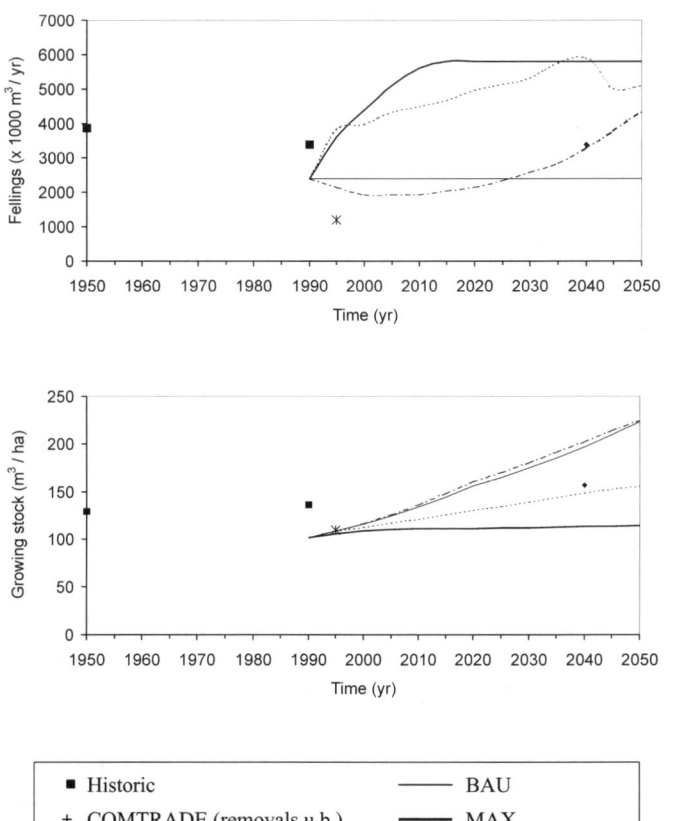

Fig. 5.10. Projected total national fellings (top) and growing stock per ha (bottom) in Bosnian forests under the four scenarios until 2050.

certain extent. That is, a rather arbitrary point of 150 m³/ha was chosen as the point where the relative growth rate is reduced and where the natural mortality increases. Still the relative growth rate was initiated in such a way that it matched the statistics, and may thus be reasonably accurate.

The approach of relative growth rate compared to growing stock also leads to increases in gross annual increment in those scenarios where growing stock continues to increase. By 2050 increments amount to between 2.8 and 4.5 m³/ha/yr for the MAX and the MTF

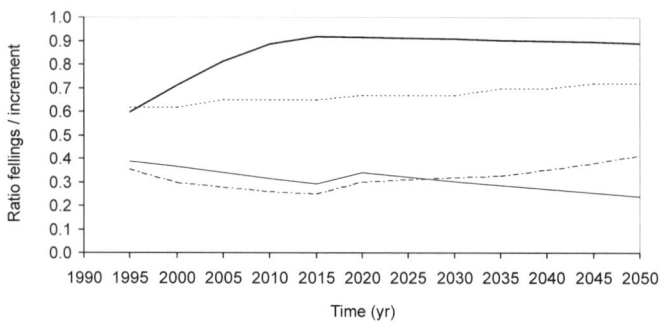

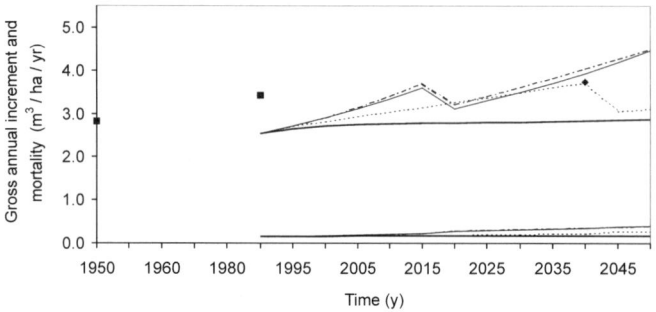

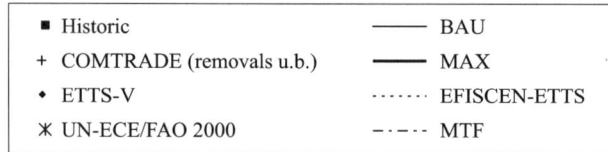

Fig. 5.11. Ratio of projected fellings/increment in Bosnian forests under the four scenarios until 2050 (top) and projected net annual increment and mortality (bottom).

scenarios, respectively. The point where the relative growth rate is reduced is clearly visible in Figure 5.8, where sudden drops in gross annual increment can be seen at points where the growing stock increases to over 150 m^3/ha. In reality, these drops will, of course, be much more gradual.

The forests of Bosnia and Herzegovina could follow the projected fellings under the EFISCEN-ETTS scenario with a required annual felling of 71% of the increment by 2050. The fellings drop slightly after 2040 because the increment is also less at that time. The output of EFISCEN-ETTS follows the output of ETTS-V very well, mainly because a comparable approach was used here.

If fellings stay at the level of the 1990s (BAU), growing stocks may increase to 223 m^3/ha. Mortality rates may then amount to 0.4 m^3/ha/yr. By 2050 growing stocks under the EFISCEN-ETTS and the MTF scenarios amount to 156 and 224 m^3/ha, respectively. Overall, the results of the MTF scenario show good possibilities for recovery of the forest sector in Bosnia and Herzegovina. On these rich sites in the Mediterranean climate annual fellings of 4.3 million m^3/yr are possible; an amount that could make Bosnia and Herzegovina a considerable producer in the future.

REFERENCES

Nilsson, S., Sallnäs, O. and Duinker, P. 1992. Future forest resources of Western and Eastern Europe. International Institute for Applied Systems Analysis. The Parthenon Publishing Group. England. 496 p.

Pajuoja, H. 1995. The outlook for the European Forest Resources and roundwood supply. Geneva, Timber and Forest Discussion Papers ECE/TIM/DP/4. UN-ECE/FAO. Geneva. 59 p.

Päivinen, R., Nabuurs, G.J., Lioubimov, A.V. and Kuusela, K. 1999. The state, utilisation and possible future developments of Leningrad region forests. EFI Working Paper 18. European Forest Institute. 59 p.

UN-ECE/FAO 1992. The Forest Resources of the Temperate Zones, the UN-ECE/FAO 1990 Forest Resource Assessment, General Forest Information. Geneva, Switzerland. 347 p.

UN-ECE/FAO 2000. Forest Resources of Europe, CIS, North America, Australia, Japan and New Zealand. Geneva, Timber and Forest Study Papers, No 17. 445 p.

5.5. BULGARIA

Introduction

Bulgaria has a wide variety of landforms, from the Danube Plains in the north, to the mountainous areas in the southwest. With 33% of the land covered with forests, it is one of the more densely forested countries in Europe. UN-ECE/FAO (2000) reports a forest area of 3.89 million ha, of which 3.1 million ha is available for wood supply. Much of the forest area consists of deciduous high forest and coppice. A change of ownership is expected, afforestation plans exist, and new forestry laws are being written. Bulgaria is one of the poorer countries in Europe and it is trying to adapt to the free market economy.

Country specific scenario assumptions

The input data covered 3.2 million ha of forest, which is the exploitable forest area according to UN-ECE/FAO (1992). The dataset is the same as used by Nilsson et al. (1992) and distinguished six types of coniferous high forest, three types of coniferous high forest, and three types of deciduous coppice. The inventory data represent the situation of around 1985.

Business-as-usual scenario

To assume an initial felling level that could be continued throughout this scenario several statistics were available. UN-ECE/FAO (1992) gave a total annual felling level of 4.5 million m³/yr having declined from 7 million m³/yr in 1970. The COMTRADE removals u.b. data depict the lowest removals level for 1994 with 2.7 million m³/yr and rising afterwards. UN-ECE/FAO (2000) report a fellings level of 4.8 million m³/yr for 1995, of which 3.1 million m³/yr is from broadleaved species. An annual fellings level of 4.5 million m³/yr was assumed in this scenario throughout the simulation period. In high forest, 30% of the fellings comes from thinnings, in coppice this is 15%. No forest area expansion was assumed here.

EFISCEN-ETTS scenario

In Bulgaria ETTS-V foresaw a rapid increase in fellings from the 4.7 million m³/yr in 1990 to 8 million m³/yr in 2040. This scenario attempted to follow this trend. There was also an area expansion of 260 000 ha by 2040. In high forest, 30% of the fellings comes from thinnings, in coppice 15% comes from thinnings.

Maximum sustainable production scenario

The fellings were increased in 15 years time to a level under which the growing stock remained approximately stable, i.e. a fellings level of 12.4 million m³/yr.

Multi-functional scenario

An initial annual fellings level of 4.5 million m³/yr was assumed. It was assumed that the economy of Bulgaria will recover, and thus the fellings will increase according to EFISCEN-ETTS scenario (around 1% per year). Furthermore, it was assumed that the policy to harvest less from final cut, will work and that by the year 2040, 60% of total fellings comes from thinnings. It was also assumed that the coppice system will be abandoned gradually, and therefore, that the fellings from coppice are reduced by 25% by 2050. The area initially over 140 years of age of high forest of oak, beech and other broadleaved species was set aside. Through ageing this can increase during the simulation. Afforestation was assumed to take place as foreseen by ETTS-V.

Results for Bulgaria

The projection of fellings and the historical data, clearly show the recent decline in economic activity in Bulgaria. The projections for 3.2 million ha of forests show that a maximum sustainable production of 11.8 million m³/yr may be achieved. This may result in a stable stocking of around 108 m³/ha, and natural mortality rates of some 5.5% of the gross annual increment.

The felling recovery as foreseen by the EFISCEN-ETTS and MTF scenarios seems rather optimistic given the present low level of activity. However, in the long run, they may appear to be realistic where Bulgaria has good export possibilities through the Black Sea. A realistic fellings level may amount to some 7.3 million m³/yr. If fellings stay at the level of the 1990s (BAU), growing stocks may increase to 194

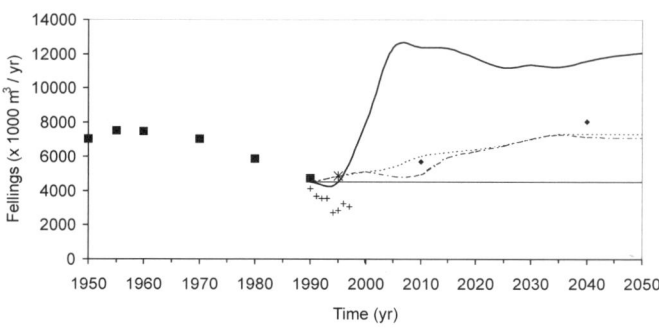

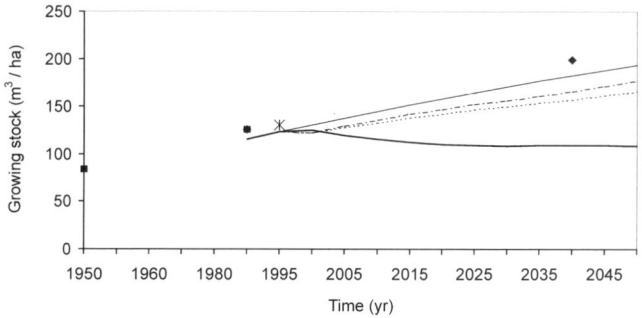

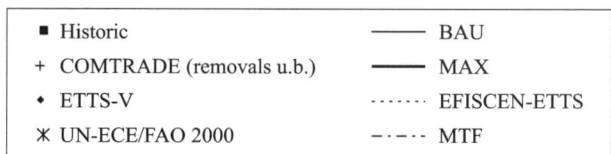

Fig. 5.12. Projected total national fellings (top) and growing stock per ha (bottom) in Bulgarian forests under the four scenarios until 2050.

m^3/ha. Mortality rates may then amount to 15% (1.3 million m^3 dead wood per year). By 2050 growing stocks under the EFISCEN-ETTS and the MTF scenarios amount to 166 and 177 m^3/ha, respectively.

In the initial age class distribution the area of coppice is clearly visible. Under the MAX scenario, this age class distribution is kept approximately the same, although the area of young forest will also consist of high forest by 2050. The large area of young forest under the MAX scenario leads to the highest gross annual increment: from 3.5 in 1985 to 3.95 m^3/ha/yr in 2050. The MTF scenario also leads

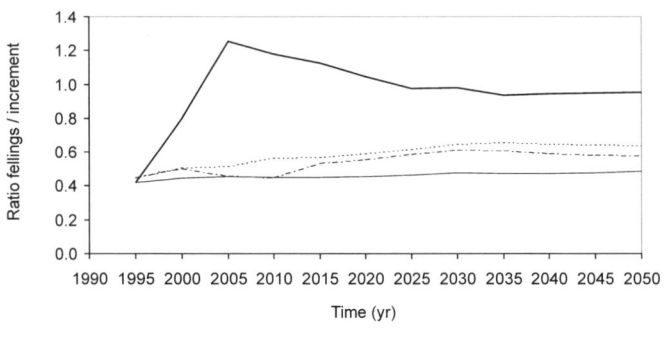

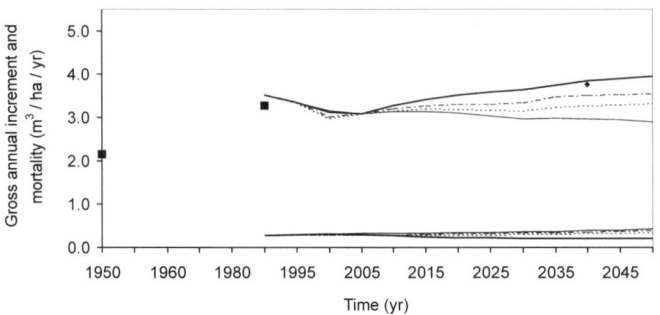

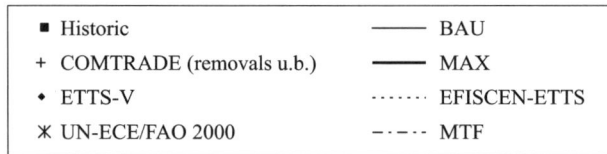

Fig. 5.13. Ratio of projected fellings/increment in Bulgarian forests under the four scenarios until 2050 (top) and projected net annual increment and mortality (bottom).

to maintenance of the gross annual increment at the level of 1985: 3.55 m³/ha/yr in 2050. This is caused by the large share of thinnings and its associated regrowth.

The BAU and MTF scenarios lead to a major shift in age class distribution, where the outgrown coppice is clearly visible by 2050. By 2050, an area of 52 000 ha is projected to be newly established as set-aside areas in the MTF scenario. This does not lead to problems in finding the required fellings.

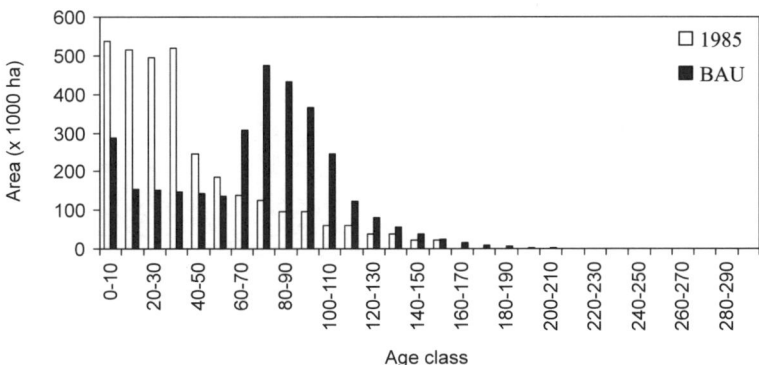

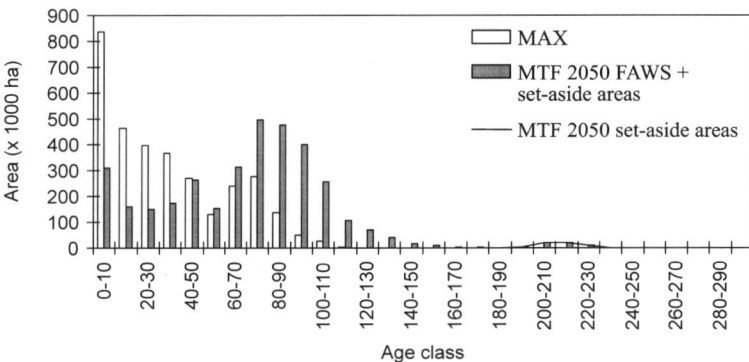

Fig. 5.14. The age class distribution of the Forest Available for Wood Supply (FAWS) in the initial situation and for 2050 under the BAU scenario (top). The age class distribution of FAWS for 2050 under the MAX scenario, the MTF scenario including set-aside areas and the MTF scenario with only set-aside areas (bottom).

REFERENCES

Nilsson, S., Sallnäs, O. and Duinker, P. 1992. Future forest resources of Western and Eastern Europe. International Institute for Applied Systems Analysis. The Parthenon Publishing Group. England. 496 p.
UN-ECE/FAO 1992. The Forest Resources of the Temperate Zones, the UN-ECE/FAO 1990 Forest Resource Assessment, General Forest Information. Geneva, Switzerland. 347 p.
UN-ECE/FAO 2000. Forest Resources of Europe, CIS, North America, Australia, Japan and New Zealand. Geneva, Timber and Forest Study Papers, No 17. 445 p.

5.6. CROATIA

In co-operation with Ms. Jela Bilandzija

Introduction

Croatia has a large variation in topography from the Adriatic Coast, to the mountainous regions of the Dinaric Alps. On the fertile, medium elevation sites, a rich deciduous forest can develop. Of the 1.77 million ha of forest as reported by UN-ECE/FA (2000), 75% is publicly owned. A considerable proportion of the forest (404 000 ha) is managed in a coppice system. Since 1991, Croatia has gone through an economic recession from which it is slowly recovering. However, the GDP per capita is amongst the lowest in Europe.

Country specific scenario assumptions

The data underlying the projections covered 1.44 million ha of forest in Croatia, this is the forest available for wood supply as given by UN-ECE/FAO (1992). Since a new inventory was not available in 1997, the initial data are the same as used by Nilsson et al. (1992) and represent the situation of around 1985. Nowadays, UN-ECE/FAO (2000) reports a forest area of 1.77 million ha for Croatia, and an increment that is almost 1 m^3/ha/yr higher than in the EEFR database. Therefore, the results presented here have to be regarded with care. The data were distinguished by four species groups and two owner classes.

Business-as-usual scenario

To assume an initial felling level that could be continued throughout this scenario several statistics were consulted: UN-ECE/FAO (2000) reports annual fellings of 4.6 million m^3/yr, of which 0.7 million m^3/yr is from coniferous species. However, Pajuoja (1995) in ETTS-V used a level of 2.97 million m^3/yr to initialise his projection. The COMTRADE removals data give a steady increase in removals since 1992: from 2.0 to 3.1 million m^3/yr in 1997. Taking into account that our data only cover 73% of the forest area, our projections were initialised with fellings of 2.2 million m^3/yr, divided between high forest and coppice according to their share in total area. In the high forest, 30% was assumed to come from thinnings, and in the coppice system this was 25%. It was assumed that the area of forest will stay at the 1990s level.

EFISCEN-ETTS scenario

ETTS-V foresaw a fast increase in fellings in Croatia, from 2.97 million m^3/yr in 1990 to 5.8 million m^3/yr in 2040. Although Pajuoja (1995) stated that these may be overestimates, the increase in projected fellings was followed. The forest area expansion of 71 000 ha between 1990 and 2020 was also followed. In the high forest, 30% was assumed to come from thinnings, and in the coppice system this was 25%.

Maximum sustainable production scenario

The fellings were increased rapidly to arrive at a point where fellings equal increment; i.e. 4.2 million m^3/yr. This was done over a period of 25 years. However, in the case of Croatia it meant that the fellings in coniferous species declined from 0.34 to 0.1 million m^3/yr.

Multi-functional scenario

The initial annual fellings of the BAU scenario was applied. It was then assumed that the economic recovery of Croatia will continue, i.e. total fellings will increase by 2% per year for the next 25 years, and then stabilise. However, it meant that the fellings in coniferous species had to decline from 0.34 to 0.1 million m^3/yr to arrive at a sustainable level. The share from thinnings gradually increased to 60% from total fellings, and it was assumed that the coppice system will be abandoned in the future and thus some fellings from coppice were transferred to the high forest. Species distribution was kept as it was in 1985. The area initially over 120 years of deciduous high forest and mixed high forest was set aside. Through ageing this can increase during the simulation. Afforestation was assumed to take place as foreseen by ETTS-V.

Results for Croatia

Since the gross annual increment applied in the current projection is almost 1 m^3/ha/yr lower than that which is reported more recently, all results have to be regarded with care. The projections for 1.44 million ha of forests show that a maximum sustainable production of

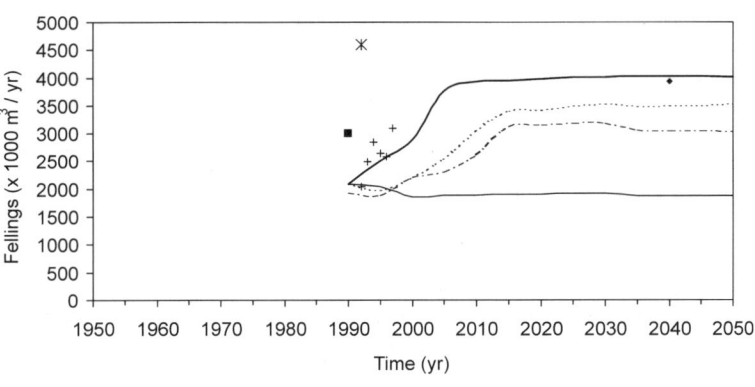

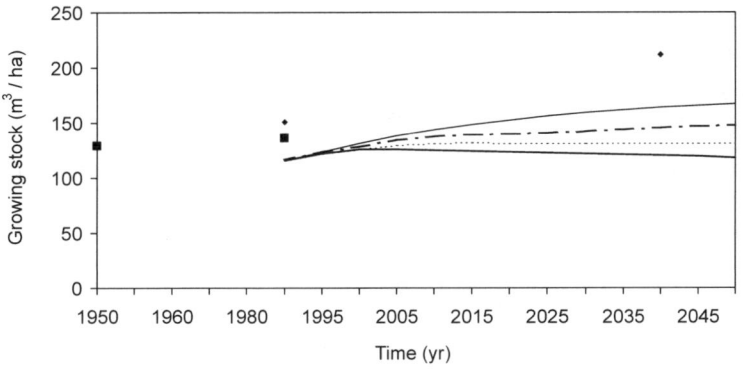

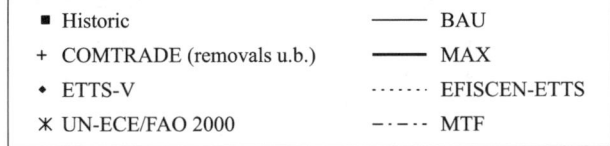

Fig. 5.15. Projected total national fellings (top) and growing stock per ha (bottom) in Croatian forests under the four scenarios until 2050.

4 million m^3/yr may be achieved. This may result in a stable stocking of around 120 m^3/ha, and natural mortality rates of some 10% of the gross annual increment.

Due to ageing effects, the gross annual increment declines in all scenarios to values around 2.6 m^3/ha/yr. The sharpest decline occurs in the BAU scenario to a gross annual increment of 2.0 m^3/ha/yr,

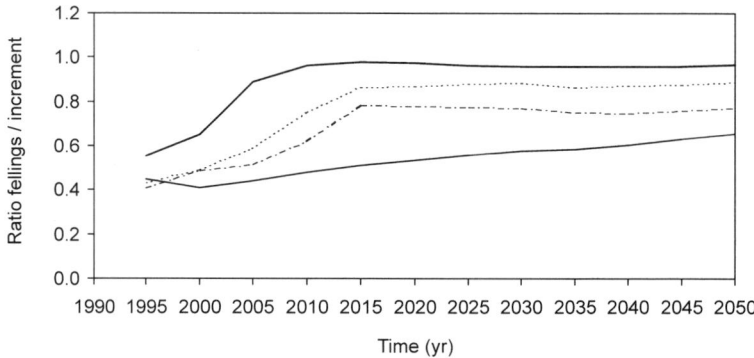

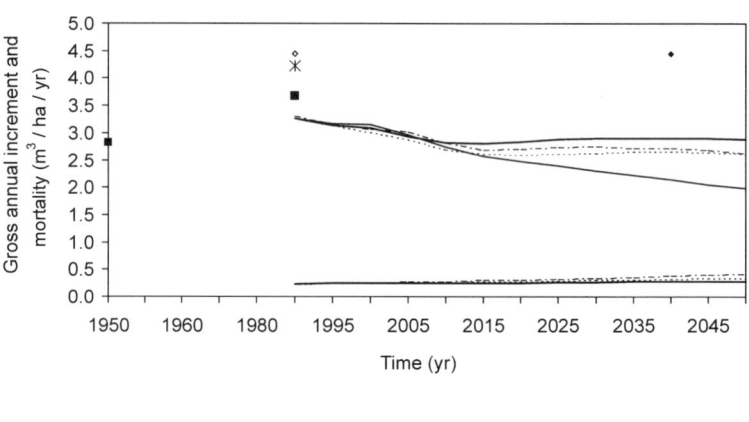

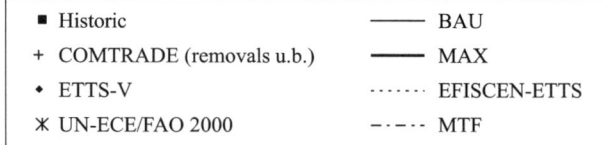

Fig. 5.16. Ratio of projected fellings/increment in Croatian forests under the four scenarios until 2050 (top) and projected net annual increment and mortality (bottom).

because in that scenario the growing stock increases from 116 m^3/ha in 1990 to 167 m^3/ha in 2050.

The COMTRADE data already showed an increasing trend in fellings in the 1990s, and thus the fast increase of 2% per year until 2025 in the EFISCEN-ETTS and MTF may be realistic. These scenarios arrive at total fellings of 3.5 million m^3/yr and 3 million

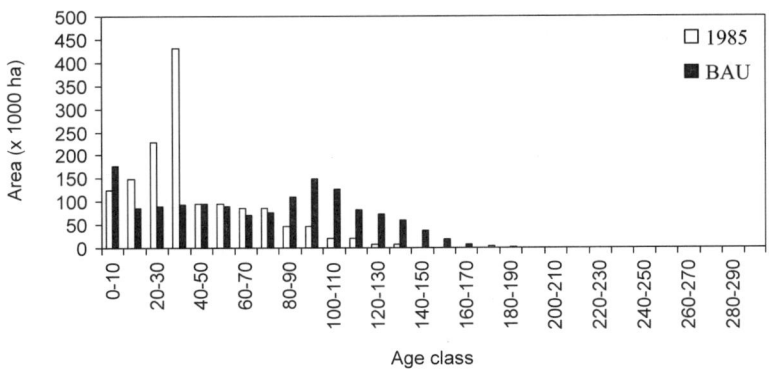

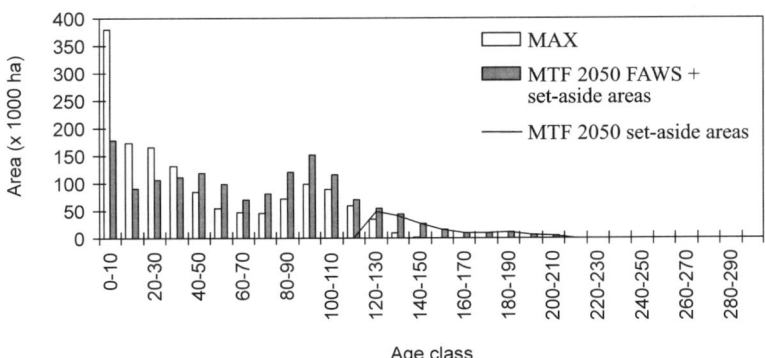

Fig. 5.17. The age class distribution of the Forest Available for Wood Supply (FAWS) in the initial situation and for 2050 under the BAU scenario (top). The age class distribution of FAWS for 2050 under the MAX scenario, the MTF scenario including set-aside areas and the MTF scenario with only set-aside areas (bottom).

m^3/yr, respectively. It is slightly lower in the MTF because fellings were reduced for the coniferous species. By 2050 growing stocks under the EFISCEN-ETTS and the MTF scenario amount to 131 and 147 m^3/ha, respectively. By 2050, the natural mortality in the MTF scenario amounts to 16% of the increment.

In the initial age class distribution the area of coppice is clearly visible. Under the MAX scenario this age class distribution is transferred to younger classes. The BAU and MTF scenarios lead to a considerable ageing of the forest, where the outgrown coppice is still visible by 2050. By 2050, an area of 171 000 ha is projected to be newly established as set-aside areas in the MTF scenario. This does not lead to problems in finding the required fellings.

REFERENCES

Nilsson, S., Sallnäs, O. and Duinker, P. 1992. Future forest resources of Western and Eastern Europe. International Institute for Applied Systems Analysis. The Parthenon Publishing Group. England. 496 p.

Pajuoja, H. 1995. The outlook for the European Forest Resources and roundwood supply. Geneva, Timber and Forest Discussion Papers. ECE/TIM/DP/4 UN-ECE/FAO. Geneva. 59 p.

UN-ECE/FAO 1992. The Forest Resources of the Temperate Zones, the UN-ECE/FAO 1990 Forest Resource Assessment, General Forest Information. Geneva, Switzerland. 347 p.

UN-ECE/FAO 2000. Forest Resources of Europe, CIS, North America, Australia, Japan and New Zealand. Geneva, Timber and Forest Study Papers, No 17. 445 p.

5.7. CZECH REPUBLIC

In co-operation with Miloš Kraus

Introduction

With 2.6 million ha of forest, the Czech Republic is one of the more densely forested countries in Europe (34% forest cover). The forest is evenly distributed over the country, but is mainly found at middle to higher elevations. The natural tree species composition was dominated by beech, but today Norway spruce covers 55% of the forest. Norway spruce has been very susceptible to health problems in the 1980s. In 1994, 67% of fellings consisted of salvaged wood, but this had dropped to 27% in 1997 (Ministry of Agriculture 1998). The felling/increment ratio is high at about 80%. Forest policy aims at a more nature-oriented forest management with a natural species composition (Kupka 1998). In the early 1990s, the Czech Republic seemed to be the first Central European country that economically benefited from the change to a market economy. However, economic growth has slowed down somewhat in the second half of the 1990s.

Country specific scenario assumptions

The data underlying the projections covered 2.45 million ha of forest in the Czech Republic, this is 96% of the forest available for wood supply as given by UN-ECE/FAO (1992) and excludes some tree species of minor importance. The data, representing the inventory cycle of 1983–1993, were distinguished by two species groups and seven regions. The Czech Republic reported underbark (u.b.) data. These data were converted to overbark (o.b.) data by multiplying by a factor 1.14 for forests older than or equal to 55 years, and by a factor 1.18 for forests younger than 55 years.

Business-as-usual scenario

The different sources for felling statistics in the Czech Republic agree to a large degree. UN-ECE/FAO (2000) reports annual fellings of 16.3 million m³/yr, of which 14.9 million m³/yr is from coniferous species. Pajuoja (1995) in ETTS-V used a level of 14 million m³/yr to initialise his projection. The COMTRADE removals data give a steady increase in removals from 10.4 million m³/yr in 1993 to 13.5 million m³/yr in 1997. Taking into account that our data cover 96% of the forest area, our projections were initialised with fellings of 14 million m³/yr. Some 25% was assumed to come from thinnings. It was assumed that the area of forests will stay at the 1990s level. It was assumed that the present species composition would not change.

EFISCEN-ETTS scenario

ETTS-V foresaw a slight decline in fellings in the Czech Republic, from 14 million m³/yr in 1990 to 13.3 million m³/yr in 2040. This was followed here. The forest area expansion of 71 000 ha between 1990 and 2000 was also followed. Some 25% of total fellings was assumed to come from thinnings.

Maximum sustainable production scenario

Conifer fellings were decreased when compared with the initial year, and in deciduous forests fellings were increased to arrive at a total felling level of 12.8 million m³/yr by 2015.

Multi-functional scenario

The initial annual fellings of the BAU scenario was applied. However, the MAX scenario indicated that Czech fellings in coniferous forest should decline from present levels to arrive at a sustainable level. This was also done for the MTF scenario, but even more strongly with an annual decline in fellings of 1% per year until 2020. Fellings in deciduous forests were slightly increased by 0.8% per year until 2020. The share from thinning gradually increased to 40% from total fellings. For harmonisation reasons with other countries, species distribution was kept as it was in 1990. The area of deciduous forest that was initially over 110 years was set aside. Through ageing this can increase during the simulation. Afforestation was assumed to take place as foreseen by ETTS-V.

Results for the Czech Republic

The present levels of fellings in the projections for the Czech Republic are at a high level. To arrive at sustainable levels, the fellings were reduced in coniferous forests in the MAX scenario. The projections for 2.45 million ha of forests then showed a maximum sustainable production of 12.6 million m³/yr. This may result in a stable stocking

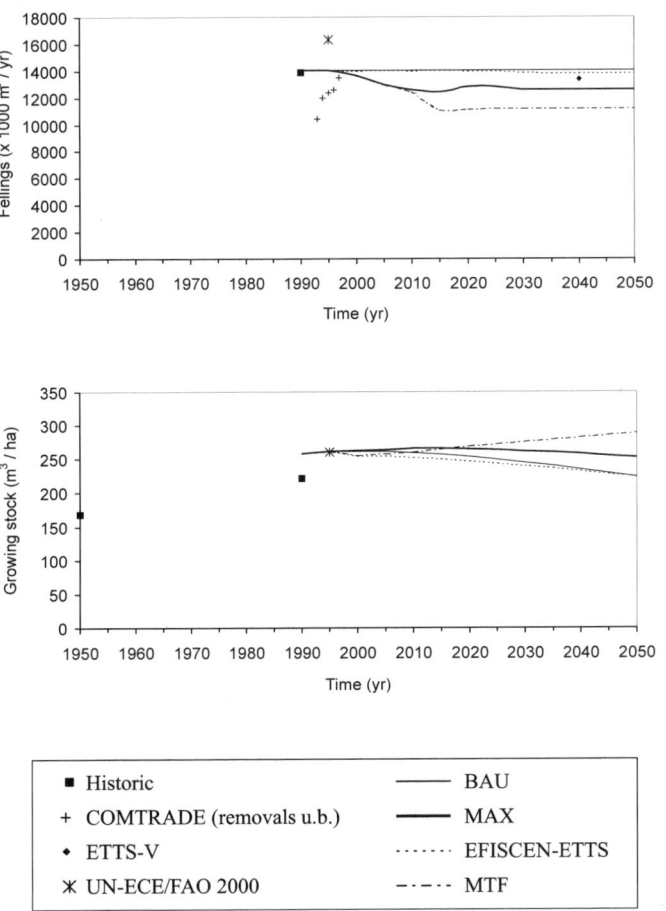

Fig. 5.18. Projected total national fellings (top) and growing stock per ha (bottom) in Czech forests under the four scenarios until 2050.

of around 252 m³/ha, and natural mortality rates rising from 10% in 1990, to 14% of the gross annual increment in 2050. The 10% in 1990 is in accordance with UN-ECE/FAO (2000).

The COMTRADE data showed an increasing trend in fellings in the 1990s, but as the results showed, that trend may have to be curbed to a felling level suggested by the MTF scenario: some 11 million m³/yr. By 2050 growing stocks under the MTF scenario amount to 289 m³/ha. By 2050, the natural mortality in the MTF scenario

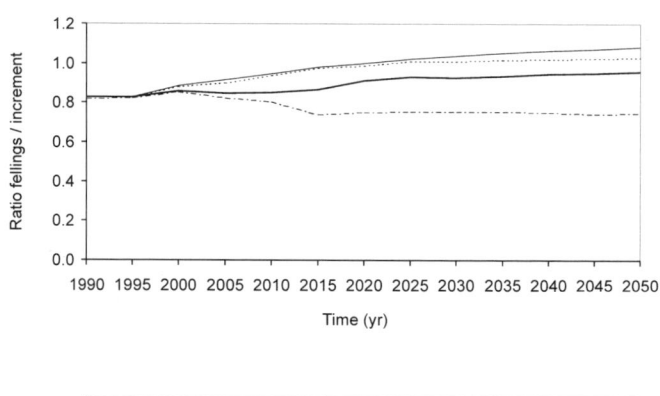

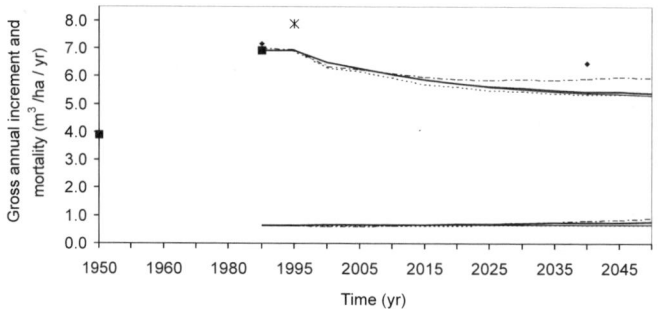

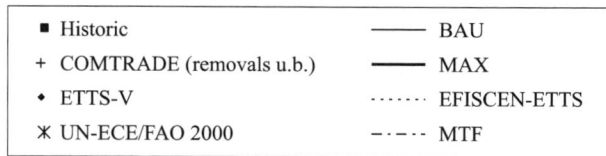

Fig. 5.19. Ratio of projected fellings/increment in Czech forests under the four scenarios until 2050 (top) and projected net annual increment and mortality (bottom).

amounts to 15% of the increment. Under the EFISCEN-ETTS and the BAU scenarios, growing stocks amount to 224 m^3/ha in 2050.

The initial age class distribution shows some concentration of forest area in the classes 50–90 years. In most scenarios that concentration is still there in 2050; mainly because of the prerequisite of sustainable felling levels. Otherwise, the MAX scenario would have pushed the age class distribution back further.

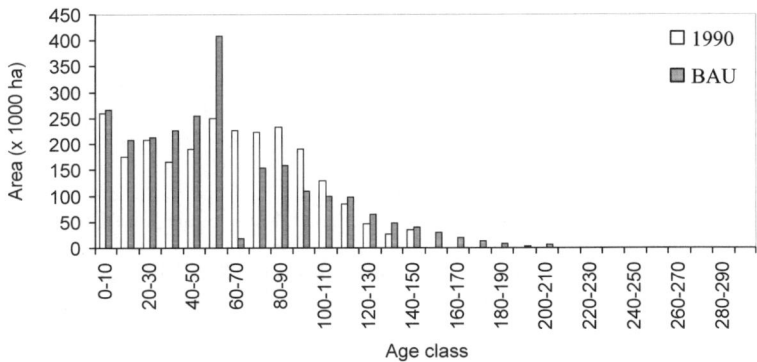

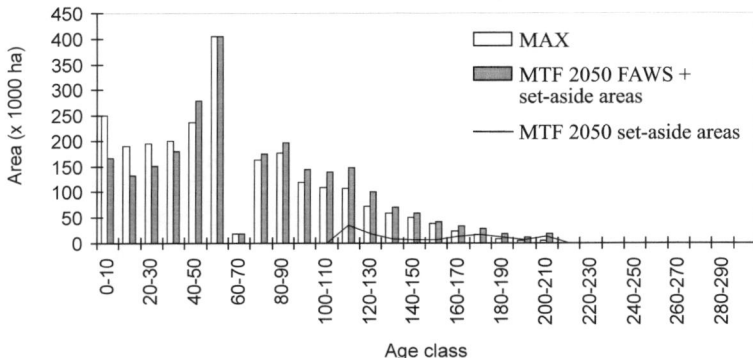

Fig. 5.20. The age class distribution of the Forest Available for Wood Supply (FAWS) in the initial situation and for 2050 under the BAU scenario (top). The age class distribution of FAWS for 2050 under the MAX scenario, the MTF scenario including set-aside areas and the MTF scenario with only set-aside areas (bottom).

Note that the clear dip in area in the class 60–70 years in the bottom graph of Figure 5.14 is the result of an initialisation problem for the clear cut area in 1990.

By 2050, an area of 138 000 ha is projected to be newly established as set-aside areas in the MTF scenario. This does not lead to problems in finding the required fellings.

The gross annual increment declines in three scenarios to values of around 5.3 m^3/ha/yr due to the occurrence of ageing. The exception is the MTF scenario with a gross annual increment of 6.0 m^3/ha/yr in 2050. This is a result of regrowth from thinning.

REFERENCES

Kupka, I. 1998. Simulation of species composition changes in the Czech forests. In: Nabuurs, G.J., Nuutinen, T., Bartelink, H. and Korhonen, M. (eds.), Forest scenario modelling for ecosystem management at landscape level. EFI Proceedings 19. European Forest Institute. Pp. 277–281.

Ministry of Agriculture of the Czech Republic 1998. Report on Forestry in the Czech Republic in 1997. Compiled by Forest Management Institute, Brandys and Labem. 173 p.

Pajuoja, H. 1995. The outlook for the European Forest Resources and roundwood supply. Geneva, Timber and Forest Discussion Papers ECE/TIM/DP/4. UN-ECE/FAO. Geneva. 59 p.

UN-ECE/FAO 1992. The Forest Resources of the Temperate Zones, the UN-ECE/FAO 1990 Forest Resource Assessment, General Forest Information. Geneva, Switzerland. 347 p.

UN-ECE/FAO 2000. Forest Resources of Europe, CIS, North America, Australia, Japan and New Zealand. Geneva, Timber and Forest Study Papers, No 17. 445 p.

Further information

Ministry of Agriculture 2001. Report on the state of forests and forestry in the Czech Republic by 31 Dec 2000. Prague. 124 p.

5.8. DENMARK

In co-operation with Dr. Peter Munk-Plum and Dr. Kim Dralle

Introduction

Denmark has a low forest cover (10%). Almost half of the 445 000 ha of forest cover are privately owned, mostly as very small holdings (UN-ECE/FAO 2000, Dragsted 1999). Some 16% of the forest consists of beech, the naturally occurring tree species. Forest policies and management aim at a more nature-oriented management, but also at a higher self-sufficiency ratio, which should be achieved by forest area expansion. Self-sufficiency in wood consumption is low, and the contribution to GDP of the forest sector is less than 0.1%. The forests are especially important as nature conservation areas and recreational spaces.

Country specific scenario assumptions

The data underlying the projections covered 0.442 million ha of forest in Denmark, this is 98% of the forest area as given by UN-ECE/FAO (2000). The data represent the state of the forest as of 1990 (Miljoministeriet 1994). For the present study the data were distinguished by five site classes and seven species.

Business-as-usual scenario

Statistics on fellings for Denmark agree closely with each other. UN-ECE/FAO (2000) reports annual fellings of 2.2 million m^3/yr, of which 1.5 million m^3/yr is from coniferous species. Pajuoja (1995) in ETTS-V used a level of 2.3 million m^3/yr to initialise his projection. The COMTRADE removals data also give a very stable fellings level: 2.3 million m^3/yr from 1993 to 1997. The projections were thus initialised with fellings of 2.3 million m^3/yr. For coniferous species, 25% of total fellings was assumed to come from thinnings; for deciduous species this was 35%. It was assumed that the area of forest will stay at the 1990s level and that there is no change in species composition.

EFISCEN-ETTS scenario

ETTS-V foresaw a rather strong increase in fellings in Denmark, from 2.3 million m^3/yr in 1990 to 4.5 million m^3/yr in 2040. This was partly based on a strong forest area expansion policy, and was followed here for as long as the desired fellings were found. The forest area expansion was projected to be 248 000 ha. For coniferous species, 25% of total fellings was assumed to come from thinnings; for deciduous species this was 35%.

Maximum sustainable production scenario

The fellings were quickly increased to arrive at a point where fellings equal increment; a level of 4 million m^3/yr by 2010.

Multi-functional scenario

The initial annual fellings of the BAU scenario was applied, and then the EFISCEN-ETTS scenario was followed for fellings. This scenario was assumed to be realistic given the policy of achieving a higher self-sufficiency in wood consumption. The rotation lengths were prolonged by 20 years. The share from thinnings gradually increased to 40% from total fellings. The area initially over 70 years old of oak and beech forest was set aside. This is a rather tight nature conservation constraint, resulting in an area of 56 000 ha being set aside in the initial year. Through ageing this can increase during the simulation. Afforestation was assumed to take place at half the rate as foreseen by ETTS-V.

Results for Denmark

The most remarkable result for Denmark is that even though an attempt was made to require fellings according to ETTS-V (in EFISCEN-ETTS and MTF), these fellings could not be found by the model. Although 4.7 million m^3/yr was required in the EFISCEN-ETTS and MTF scenarios, the model 'found' only 3.6 million and 2.7 million m^3/yr, respectively. This is a result of the way the management regimes were set, and the more stringent management regimes in the MTF scenario. It is also caused by the fact that the new afforestations will only reach a harvestable age after some 70 years, and are thus not available for final felling by the end of the simulation time.

The projections for 0.442 million ha of forests show that a maximum sustainable production of 3.6 million m^3/yr may be achieved. This may

DENMARK

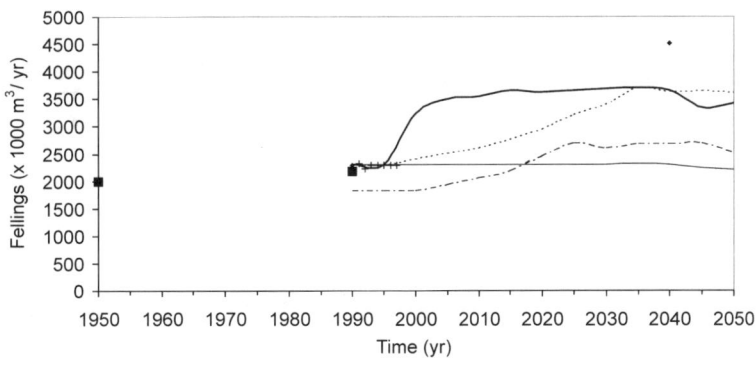

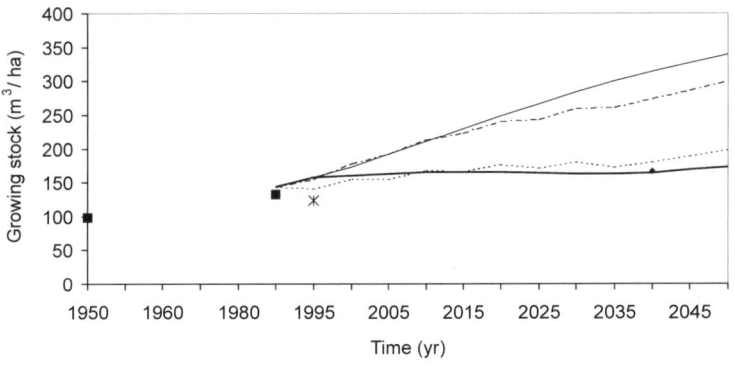

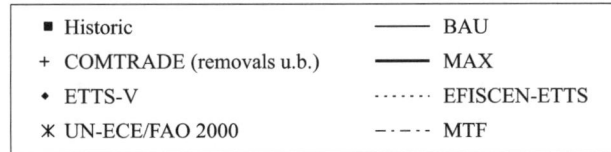

Fig. 5.21. Projected total national fellings (top) and growing stock per ha (bottom) in Danish forests under the four scenarios until 2050.

result in a stable stocking of around 174 m^3/ha, and natural mortality rates of 3% of the gross annual increment.

By 2050 growing stocks under the EFISCEN-ETTS and the MTF scenarios amount to 198 and 300 m^3/ha, respectively. So, even though the fellings cannot be found, the growing stocks increase under both scenarios. By 2050, the natural mortality in the MTF scenario amounts to 8% of the increment.

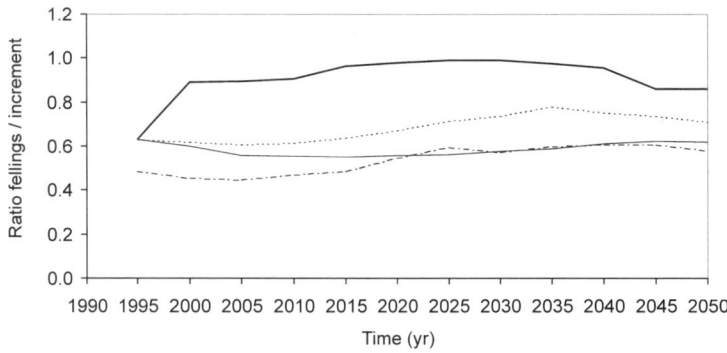
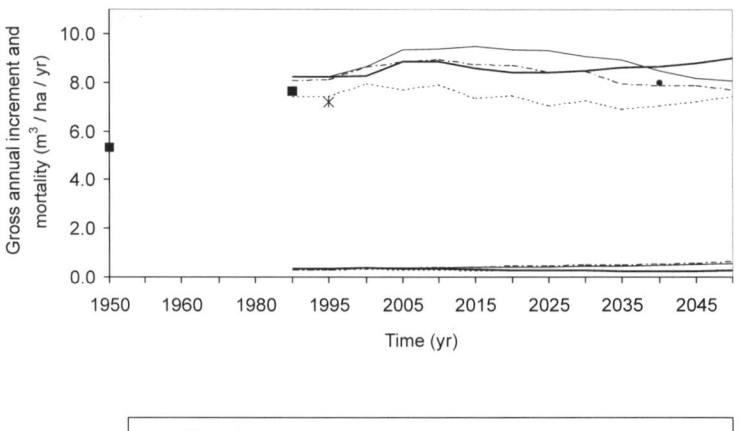
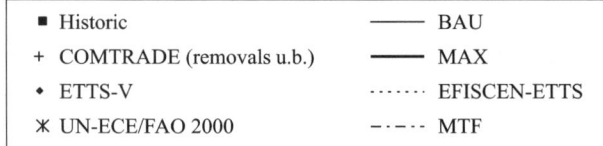

Fig. 5.22. Ratio of projected fellings/increment in Danish forests under the four scenarios until 2050 (top) and projected net annual increment and mortality (bottom).

The gross annual increment is very stable in all scenarios with values around 8 m³/ha/yr. The lowest NAI was found for the EFISCEN-ETTS scenario, because in that scenario substantial afforestations are carried out, and thus very young, slow growing forest area is added.

A clear ageing effect can be seen in the age class distributions for 2050 in the BAU and the MTF scenarios. This ageing is enhanced

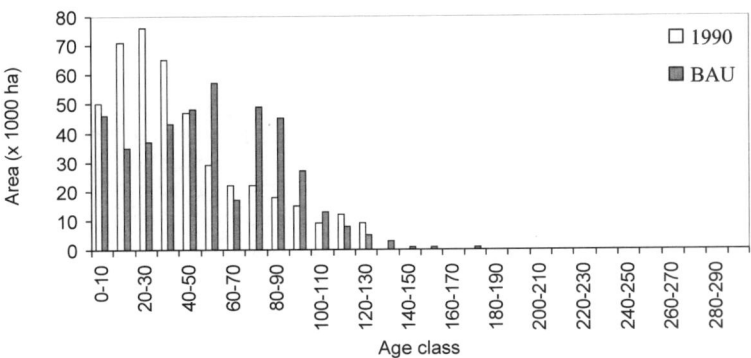

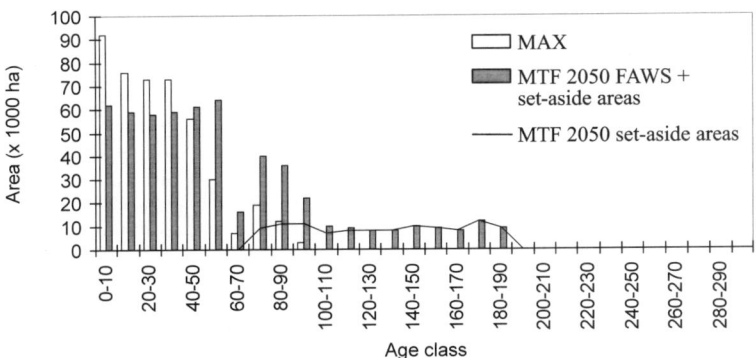

Fig. 5.23. The age class distribution of the Forest Available for Wood Supply (FAWS) in the initial situation and for 2050 under the BAU scenario (top). The age class distribution of FAWS for 2050 under the MAX scenario, the MTF scenario including set-aside areas and the MTF scenario with only set-aside areas (bottom).

through the establishment of reserves in the MTF scenario; an area of 110 000 ha is projected to be newly established as set-aside areas by 2050.

REFERENCES

Dragsted, J. 1999. Denmark. In: Pelkonen, P., Pitkänen, A., Schmidt, P., Oesten, G., Piussi, P. and Rojas, E. (eds.), Forestry in Changing Societies in Europe. Part II. SILVA Network. University Press, Joensuu, Finland. Pp. 45–59.

Miljoministeriet 1994. Skove og plantager 1990. Danmarks Statistics, Copenhagen. 131 p.

Pajuoja, H. 1995. The outlook for the European Forest Resources and roundwood supply. Geneva Timber and Forest Discussion Papers ECE/TIM/DP/4. UN-ECE/FAO. Geneva. 59 p.

UN-ECE/FAO 2000. Forest Resources of Europe, CIS, North America, Australia, Japan and New Zealand. Geneva, Timber and Forest Study Papers, No 17. 445 p.

5.9. FINLAND

In co-operation with Prof. Erkki Tomppo, Ms. Tarja Tuomainen, Dr. Tuula Nuutinen and Mr. Eric Verkaik

Introduction

With 72% forest cover, Finland is one of the largest forestry countries in Europe. Just over 60% of the 21.9 million ha of boreal forest is owned by private owners (UN-ECE/FAO 2000, Räisänen 1999). The majority of the forest consists of three species: Norway spruce, Scots pine and birch. Finland renewed its forest legislation in 1997 and prepared a national forest programme in 1999, both placing a greater emphasis on protection of biological diversity of forests more than before. This has had an impact on forest management practices. Finland is a major exporter of sawn softwood and paper. The share of forestry and forest industry in GDP amounts to 7.7%. An increasing share (now at 8.5 million m^3/yr) of the raw material (mainly birch) is imported from Russia.

Country specific scenario assumptions

The data underlying the projections covered 19.63 million ha of forest in Finland and represent the area available for wood supply according to UN-ECE/FAO (1992). The data are based on the 1986–1994 inventory cycle. The data were distinguished by two regions, eight site classes and four tree species. Since birch and aspen are usually mixed in other stands, the basic data underestimated the total growing stock for these two species. A correction was carried out for this.

Business-as-usual scenario

The statistical yearbook reported annual fellings of 55.1 million m^3/yr, of which 20% is from deciduous species (Metla 1997). UN-ECE/FAO (2000) reports annual fellings of 54.3 million m^3/yr, of which 43.5 million m^3/yr is from coniferous species. The COMTRADE removals data give an increasing trend from 43 million m^3/yr in 1993 to 51 million m^3/yr in 1997. The projections were thus initialised with fellings of 54 million m^3/yr. Some 25% of total fellings was assumed to come from thinnings. It was assumed that the area of forest will stay at the level of the 1990s. The three main species in Finnish forest are indigenous species, and it was assumed that there is no change in species composition.

EFISCEN-ETTS scenario

ETTS-V foresaw very stable felling levels in Finland; around 52 million m^3/yr. This was followed here. The forest area was assumed to stay the same, but they foresaw a strong increase in increments from 4.2 m^3/ha/yr in 1990 to 6.3 m^3/ha/yr in 2040.

Maximum sustainable production scenario

The fellings were quickly increased to arrive at a point where fellings equal increment; a level of 62 million m^3/yr by 2015.

Multi-functional scenario

The initial annual fellings of the BAU scenario was applied, and an increase of 0.5% per year in coniferous fellings was foreseen until 2020. After that it stabilises. No increase in fellings in deciduous species was foreseen. To take into account the current discussion concerning nature-oriented management, the rotation lengths were prolonged by 20 years. The share from thinnings gradually increased to 50% from total fellings. The area initially over 160 years of age of birch and aspen was set aside. Through ageing this can increase during the simulation. Afforestation was assumed to take place on marginal agricultural sites amounting to a total of 96 000 ha by 2010 and another 96 000 ha by 2030.

Results for Finland

Maybe the most remarkable result for Finland is that the trends are rather slow: the low growth rate, and the rather high fellings/increment ratio make developments in age class distribution and in growing stock rather modest. Furthermore, it is apparent that the results from the present study differ rather much from the ETTS-V results; mainly because a strong increase in increment was expected in the latter study.

The projections for the 19.63 million ha of forests show that a maximum sustainable felling level of 64.2 million m^3/yr can be achieved. This may result in a stable stocking of around 91 m^3/ha, and natural mortality rates of 5% of the gross annual increment.

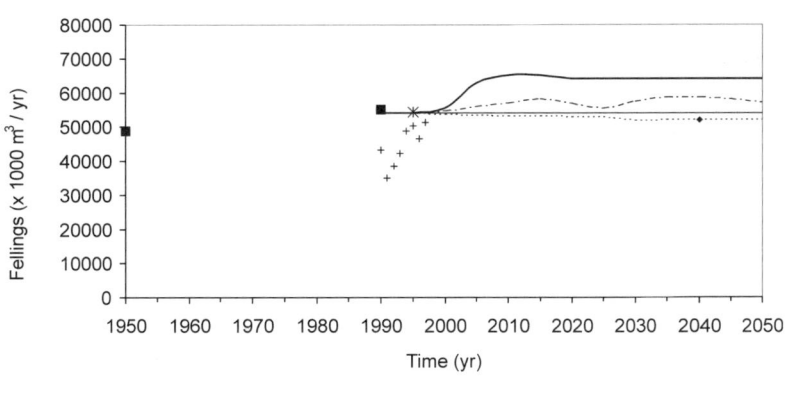

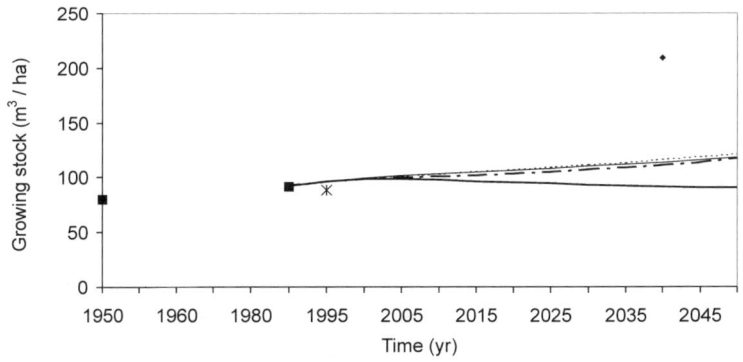

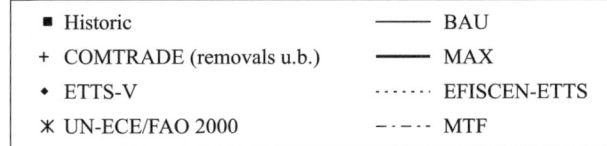

Fig. 5.24. Projected total national fellings (top) and growing stock per ha (bottom) in Finnish forests under the four scenarios until 2050.

By 2050 the average growing stock under the MTF scenario amounts to 117 m^3/ha, whereas ETTS-V estimated that it could be 209 m^3/ha. By 2050, the natural mortality in the MTF scenario amounts to 7% of the increment. The gross annual increment declines slightly in the beginning of each scenario until around 2010, and is very stable afterwards; in most scenarios it varies around 3.4 m^3/ha/yr, but it is slightly higher in the MTF scenario at 3.9 m^3/ha/yr.

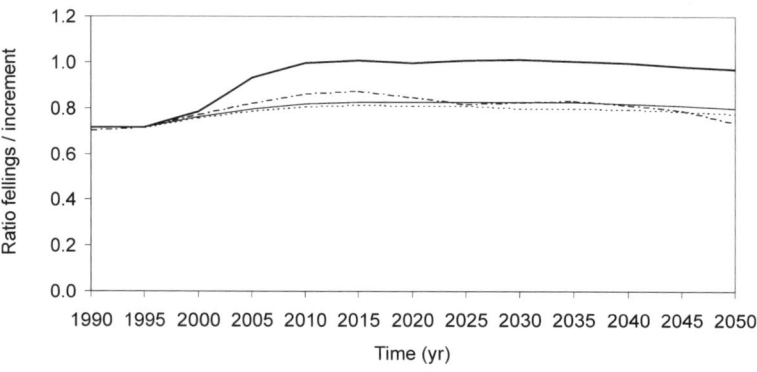

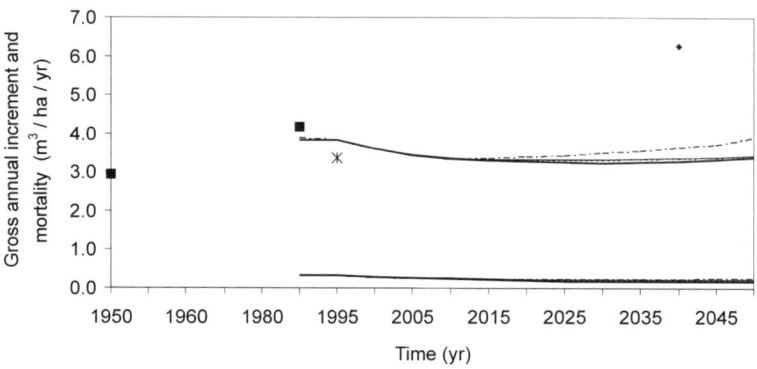

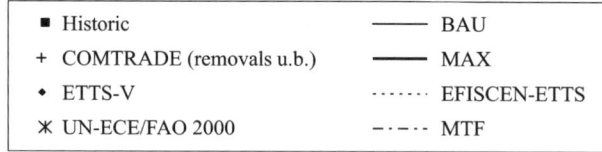

Fig. 5.25. Ratio of projected fellings/increment in Finnish forests under the four scenarios until 2050 (top) and projected net annual increment and mortality (bottom).

In the age class distributions for 2050, mainly the MAX scenario shows a shift compared with 1990; the areas are pushed back towards younger classes. The MTF clearly demonstrates the consequences of the establishment of new reserves; an area of 1.24 million ha is projected to be set aside by 2050.

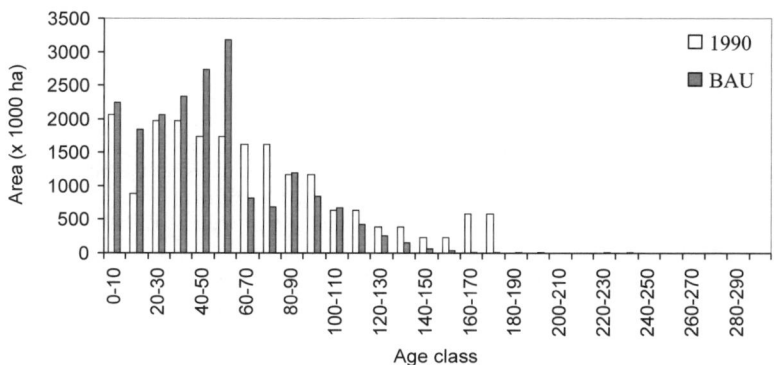

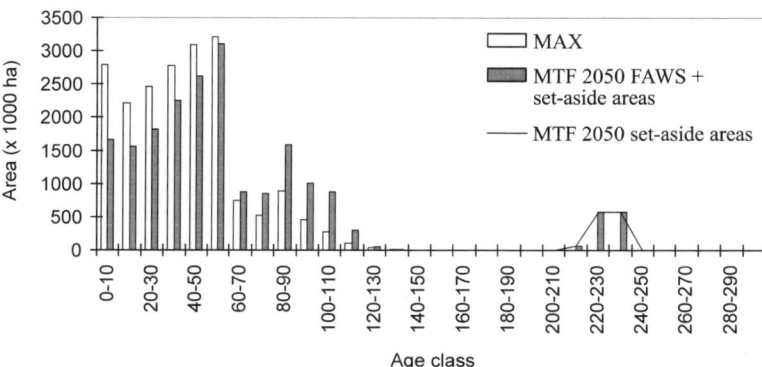

Fig. 5.26. The age class distribution of the Forest Available for Wood Supply (FAWS) in the initial situation and for 2050 under the BAU scenario (top). The age class distribution of FAWS for 2050 under the MAX scenario, the MTF scenario including set-aside areas and the MTF scenario with only set-aside areas (bottom).

REFERENCES

Räisänen, S. 1999. Finland. In: Pelkonen, P., Pitkänen, A., Schmidt, P., Oesten, G., Piussi, P. and Rojas, E. (eds.), Forestry in Changing Societies in Europe. Part II. SILVA Network. University Press, Joensuu, Finland. Pp. 61–75.

Metla 1997. Metsätilastollinen vuosikirja 1997. Finnish Statistical Yearbook of Forestry. Metla, Helsinki, Finland. 348 p.

UN-ECE/FAO 1992. The Forest Resources of the Temperate Zones, the UN-ECE/FAO 1990 Forest Resource Assessment, General Forest Information. Geneva, Switzerland. 347 p.

UN-ECE/FAO 2000. Forest Resources of Europe, CIS, North America, Australia, Japan and New Zealand. Geneva, Timber and Forest Study Papers, No 17. 445 p.

Further information

Siitonen, M. and Nuutinen, T. 1996. Timber production analyses in Finland and the MELA system. In: Päivinen, R., Roihuvuo, L. and Siitonen, M. (eds.), Large-scale forestry scenario models: experiences and requirements. EFI Proceedings 5. European Forest Institute. Pp. 89–98.

Ministry of Agriculture and Forestry 1999. Finland's national forest programme 2010. Publication 2. Helsinki.

5.10. FRANCE

In co-operation with Mr. Gérôme Pignard, Dr. Bin Guo and Mr. Yann-Nicolas Bouton

Introduction

With excellent growing conditions and 28% of its land covered with forests (15.2 million ha), France is one of the larger forest countries in Europe. Three-quarters of the forest area is privately owned. The variety of sites, from the Atlantic coast to the high Alpine and Mediterranean sites, and the very long history of forest management, give the forests of France an enormous cultural historical richness and high biodiversity. A considerable proportion of the mountainous forests is managed in the selection system (Piel et al. 1999). France is the second largest producer in Europe with some 60 million m^3 roundwood/yr, about equally distributed between coniferous and deciduous species. The country is renowned for its high quality oak wood. However, the forestry sector contributes only 1.5% to the GDP.

Country specific scenario assumptions

The data underlying the projections covered 8.4 million ha of forest by age class distributions, and 4.9 million ha by diameter class distribution. Out of the 8.4 million ha of even-aged forest, 2.1 million ha was under coppice management. Thus the two modelling approaches were used and results were aggregated afterwards. The data represent the area available for wood supply according to UN-ECE/FAO (1992). The data are based on the 1983–1993 inventory cycle. The age class data were distinguished by six regions, two owner groups, two site classes and eight tree species. The diameter class data were distinguished by six regions, three owner groups, and five tree species.

Business-as-usual scenario

UN-ECE/FAO (2000) reports a production of 60.2 million m^3 roundwood/yr, about equally distributed between coniferous and deciduous species. UN-ECE/FAO (1992) reports a production of 52 million m^3/yr. The COMTRADE removals underbark data give a slightly decreasing trend from 45 million m^3/yr in 1990 to 41 million m^3/yr in 1997. The projections were thus initialised with fellings of 52 million m^3/yr. In the high forest, 30% of total fellings was assumed to come from thinnings, for coppice this was assumed to be 20%. It was assumed that the area of forest will stay at the 1990s level without changes in species distribution.

EFISCEN-ETTS scenario

ETTS-V foresaw a gradual increase in fellings in France from 52 million m^3/yr in 1990 to 68 million m^3/yr in 2040. This was followed here. The forest area expansion of ETTS-V was followed as well, with 500 000 ha before 2000, 40 000 ha/yr from 2000 to 2010 and 10 000 ha/yr from 2010 to 2040. It was assumed that all of this afforestation will take place with regular high forest of Maritime pine, Douglas-fir, other conifers, Norway spruce/fir, and other pines. In total the simulated forest area thus increases from 13.3 million ha in the initial year to 14.5 million ha in 2050.

Maximum sustainable production scenario

The fellings were quickly increased to arrive at a point where fellings equal increment; i.e. a level of 61.8 million m^3/yr by 2010.

Multi-functional scenario

The initial annual fellings of the BAU scenario was applied, and then the idea of ETTS-V was followed, but at a slower pace. An increase of 0.8% per year in coniferous fellings was foreseen until 2030. After that it stabilises. An increase in fellings in deciduous species of 0.6% per year was foreseen until 2020. To take into account the ideas of nature-oriented management, the rotation lengths were prolonged by 20 years. The share from thinnings gradually increased to 40% from coniferous total fellings, 30% from deciduous high forest total fellings, and 20% from deciduous coppice. The area initially over 100 years of age of oak, beech and mixed oak beech high forest was set aside. Through ageing this can increase during the simulation. Afforestation was assumed to take place at half the pace of the rate of ETTS-V.

Results for France

The French results show that the current production is rather close to its biological maximum given the management regimes that keep the age class distribution rather stable: a maximum sustainable production

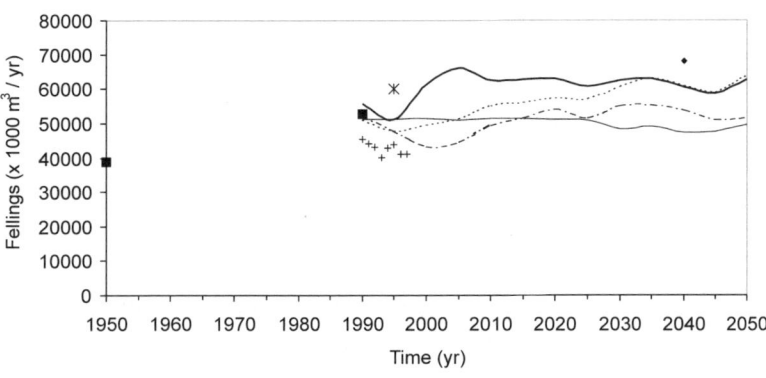

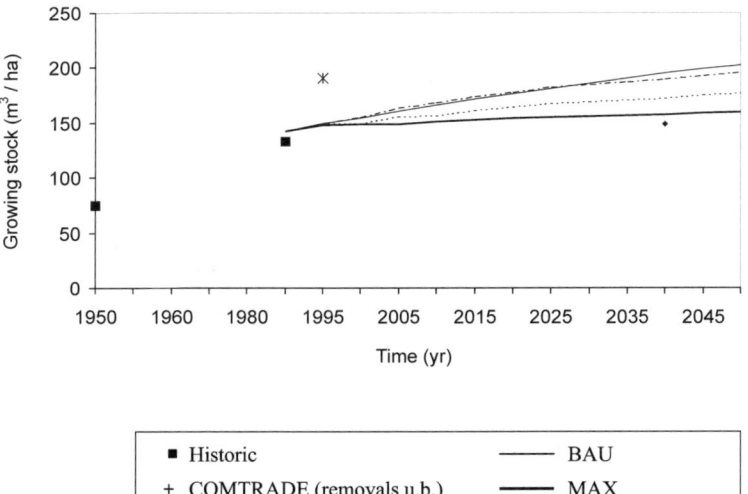

Fig. 5.27. Projected total national fellings (top) and growing stock per ha (bottom) in French forests under the four scenarios until 2050.

of 62.2 million m³/yr may be achieved, Under this level (with 92% of the increment harvested) the growing stock slightly increases to 160 m³/ha in 2050, and natural mortality amounts to 8.3% of the gross annual increment.

The rather high fellings/increment ratio makes developments in age class distribution and in growing stock rather modest in France.

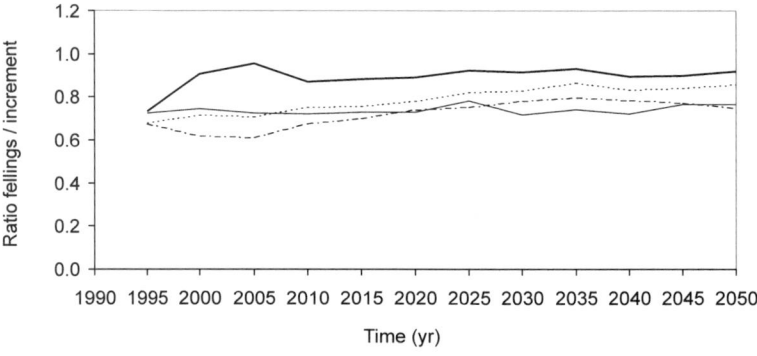
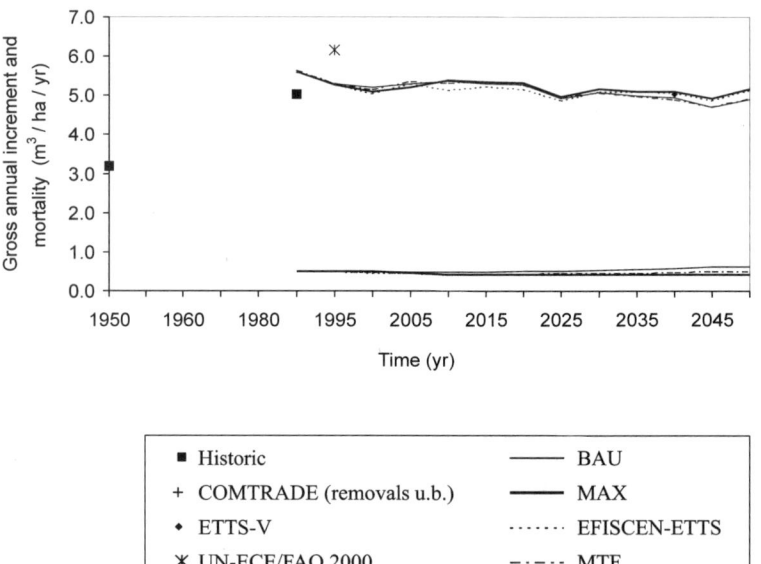

Fig. 5.28. Ratio of projected fellings/increment in French forests under the four scenarios until 2050 (top) and projected net annual increment and mortality (bottom).

Furthermore, it is apparent that the results from the present study differ from the ETTS-V results that foresaw a strong increase in fellings, and thus a much lower growing stock in 2040. The BAU, the MTF and the EFISCEN-ETTS scenarios all show difficulty in meeting the required fellings after 2035. Since the growing stock is continuously building up

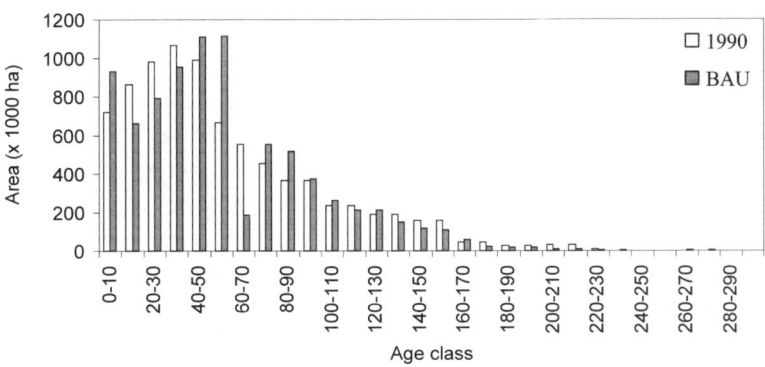

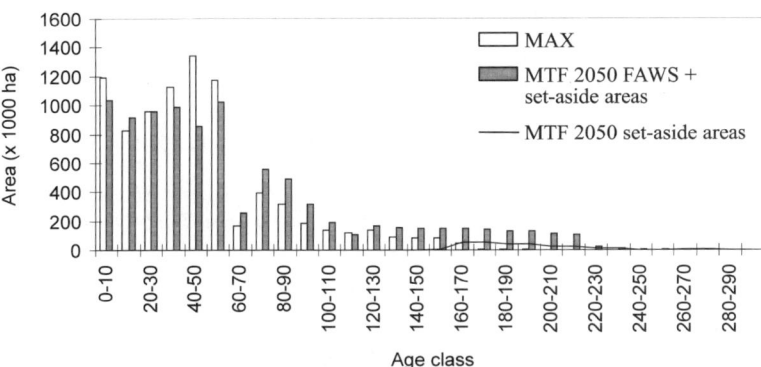

Fig. 5.29. The age class distribution of the Forest Available for Wood Supply (FAWS) in the initial situation and for 2050 under the BAU scenario (top). The age class distribution of FAWS for 2050 under the MAX scenario, the MTF scenario including set-aside areas and the MTF scenario with only set-aside areas (bottom).

in these scenarios, this may be seen as a virtual shortage; the current felling regimes cannot meet future demand.

By 2050 the average growing stock under the MTF scenario amounts to 197 m^3/ha with a mortality rate of 14.6% of the increment, or a total production of dead wood of 10 million m^3. The gross annual increment declines slightly in the beginning of each scenario until around 2005, and is stable afterwards; in all scenarios it varies around 5 m^3/ha/yr, but is slightly higher in the EFISCEN-ETTS and the MAX scenarios, around 5.2 m^3/ha/yr. This is considerably lower (and thus affects the results) than the latest statistic reported for French increment to UN-ECE: 6.15 m^3/ha/yr.

In the age class distributions for 2050, the MAX scenario shows a shift compared with 1990; the areas are pushed back towards younger classes. The MTF scenario shows the ageing of the forest and the established reserves; an area of 271 000 ha is projected to be newly established as set-aside areas by 2050. This rather modest establishment of new reserves is only part of the reason why the fellings in MTF decline after 2035.

REFERENCES

Piel, A., Costa, S. and Peyron, J.L. 1999. France. In: Pelkonen, P., Pitkänen, A., Schmidt, P., Oesten, G., Piussi, P. and Rojas, E. (eds.), Forestry in Changing Societies in Europe. Part II. SILVA Network. University Press, Joensuu, Finland. Pp. 79–98.

UN-ECE/FAO 1992. The Forest Resources of the Temperate Zones, the UN-ECE/FAO 1990 Forest Resource Assessment, General Forest Information. Geneva, Switzerland. 347 p.

UN-ECE/FAO 2000. Forest Resources of Europe, CIS, North America, Australia, Japan and New Zealand. Geneva, Timber and Forest Study Papers, No 17. 445 p.

Further information

Pignard, G. 1996. Data and models for a large scale forestry analysis in France. In: Päivinen, R., Roihuvuo, L. and Siitonen, M. (eds.), Large scale forestry scenario models: experiences and requirements. EFI Proceedings 5. European Forest Institute. Pp. 289–294.

Pignard, G. 2000. Évolution récente des forêts françaises: surface, volume surface, volume sur pied, productivité. Revue Forestière Française. 52: 27–36.

5.11. GERMANY

In co-operation with Dr. Volker Sasse and D.I. Hermann Englert

Introduction

The 10.7 million ha of forests in Germany stretch from the Northern German plains to the Central European middle mountain sites (UN-ECE/FAO 2000). On the northern plains Scots pine is the dominant species, while in the middle mountainous areas the dominant species are Norway spruce and beech. With an annual production of around 50 million m^3/yr, Germany is among the larger producers in Europe, especially of coniferous wood. Despite this, the forestry sector contributes only marginally to the GDP. The forest health discussions of the 1980s were very intense in Germany and have led to considerable pressure from society to change forest management to a more nature-oriented management. However, the unification of Germany and the budgetary problems that came along with it, pushed the society's grave concerns over environmental problems to a level of general awareness (Schraml and Winkler 1999).

Country specific scenario assumptions

The data underlying the projections covered 9.9 million ha of forest, thus covering the forests available for wood supply according to UN-ECE/FAO (1992). The data are based on the 1986–1990 and the 1993 inventory cycle. The data were distinguished by 13 regions, and nine tree species. Since then some Bundesländer have renewed their inventory, but those data were not used.

Business-as-usual scenario

To initialise the required national production, several statistics were used. UN-ECE/FAO (2000) reports a production of 48.5 million m^3 roundwood/yr, of which 77% is from conifers. UN-ECE/FAO (1992) reported a production of 42.8 million m^3/yr. ETTS-V, which was initialised also for 1990, starts with annual fellings of 68.8 million m^3/yr, but that may have been caused by the storm of that year. The storm year is also noticeable in the COMTRADE removals underbark data; they show a slightly increasing trend from 33 million m^3/yr in 1991 to 38 million m^3/yr in 1997. The projections were thus initialised with fellings of 43 million m^3/yr. Some 30% of total fellings from conifers was assumed to come from thinnings, and for deciduous species this was assumed to be 25%. It was assumed that the area of forest will stay at the 1990s level and that there is no change in species composition.

EFISCEN-ETTS scenario

ETTS-V foresaw a very stable, but also very high, level of fellings in Germany from 68.7 million m^3/yr in 1990 to 75.5 million m^3/yr in 2040. This was followed here, although we think that in ETTS-V the level of the storm year 1990 was used throughout. ETTS-V foresaw no forest area expansion.

Maximum sustainable production scenario

The fellings were quickly increased to arrive at a point where fellings equal increment; i.e. a level of 80 million m^3/yr by 2005.

Multi-functional scenario

The initial annual fellings of the BAU scenario was applied, and then it was assumed that the fellings will increase by 1% per year until 2010 to arrive at a level of 52 million m^3/yr. After that fellings stabilise. The rotation lengths were prolonged by 20 years. The share from thinnings gradually increased to 50% from total fellings. The initial set-aside areas were the areas over 160 years of oak, over 140 years of beech and over 120 years for other deciduous species. Initially this represents 176 000 ha. Through ageing this can increase during the simulation. Afforestation was assumed to take place at the pace that UN-ECE/FAO (2000) reports: 22 000 ha/yr in the period 1990–2015.

Results for Germany

The projections for the 9.9 million ha of forests show that a maximum sustainable production of 80 million m^3/yr may be achieved, which is a large increase from the present level of around 43 million m^3/yr. Thus under most scenarios the present high stocking of 266 m^3/ha increases further; by 2050 the growing stock under the MTF scenario may amount to 405 m^3/ha, and under the BAU it may amount to

GERMANY

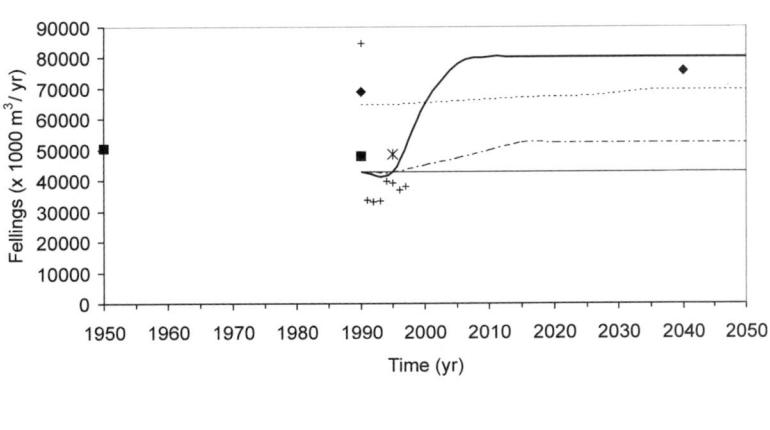

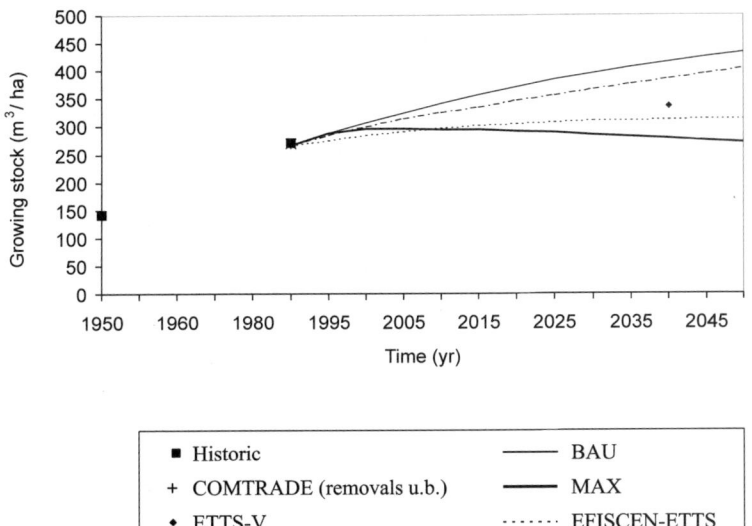

Fig. 5.30. Projected total national fellings (top) and growing stock per ha (bottom) in German forests under the four scenarios until 2050.

433 m³/ha. Natural mortality rates may then amount to 11.7 and 15.2% of the gross annual increment, respectively.

The gross annual increment declines slightly in all scenarios: for example, from 9.4 m³/ha/yr initially to 8.1 m³/ha/yr in 2050 in the MAX scenario. The decline is the strongest in the scenario where the

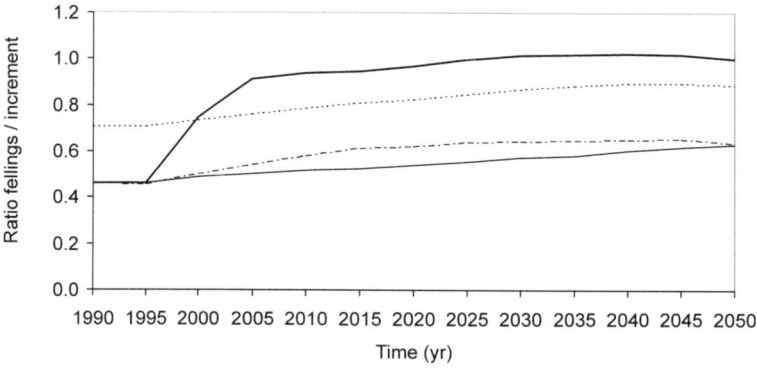

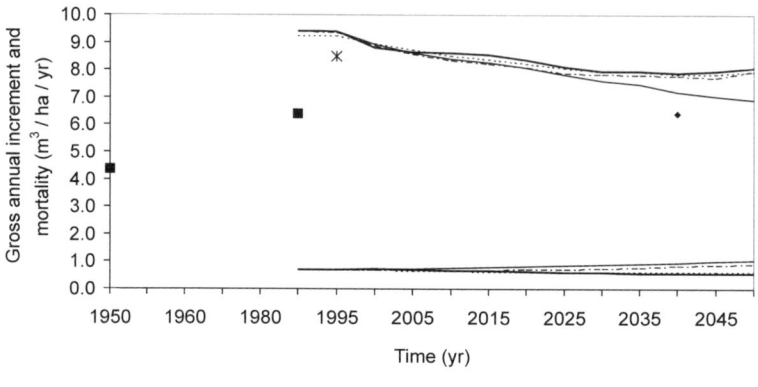

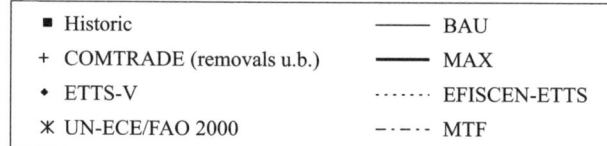

Fig. 5.31. Ratio of projected fellings/increment in German forests under the four scenarios until 2050 (top) and projected net annual increment and mortality (bottom).

build-up of growing stock is the largest: the BAU scenario with an increment of 6.9 m³/ha/yr in 2050.

The age class distributions for 2050 clearly show the ageing process because the utilisation ratio in most scenarios is around 60% only. The ageing process is somewhat less in the MTF scenario compared

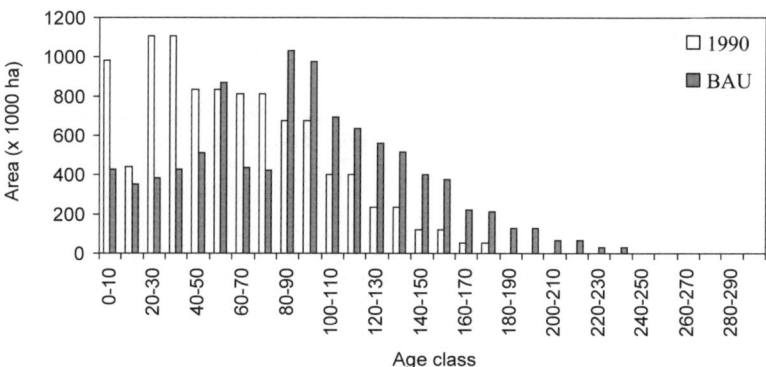

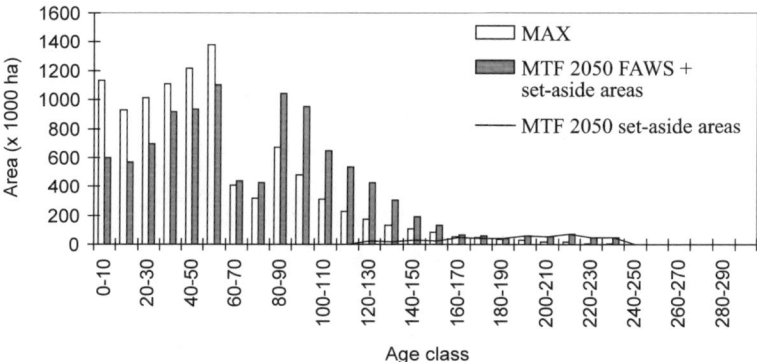

Fig. 5.32. The age class distribution of the Forest Available for Wood Supply (FAWS) in the initial situation and for 2050 under the BAU scenario (top). The age class distribution of FAWS for 2050 under the MAX scenario, the MTF scenario including set-aside areas and the MTF scenario with only set-aside areas (bottom).

with the BAU, because in the MTF scenario, 504 000 ha are projected to be newly established as set-aside areas by 2050 and the utilisation ratio is slightly higher in the MTF scenario. This leads to more intense logging in the remaining area available for wood supply.

REFERENCES

Schraml, U. and Winkel, G. 1999. Germany. In: Pelkonen, P., Pitkänen, A., Schmidt, P., Oesten, G., Piussi, P. and Rojas, E. (eds.), Forestry in Changing Societies in Europe. Part II. SILVA Network. University Press, Joensuu, Finland. Pp. 115–138.

UN-ECE/FAO 1992. The Forest Resources of the Temperate Zones, the UN-ECE/FAO 1990 Forest Resource Assessment, General Forest Information. Geneva, Switzerland. 347 p.

UN-ECE/FAO 2000. Forest Resources of Europe, CIS, North America, Australia, Japan and New Zealand. Geneva, Timber and Forest Study Papers, No 17. 445 p.

Further information

Bick, U. and Dahm, S. (eds.) 1992. Bundeswaldinventur 1986–1990. Bundesministerium für Ernährung, Landwirtschaft und Forsten. Vol. 1: Inventurbericht und Übersichtstabellen für das Bundesgebiet nach dem Gebietsstand vor dem 3.10.1990 einschliesslich Berlin (West).

Englert, H. and Sasse, V. 1994. Entwicklung des Schwachholzaufkommens in den neuen Deutschen Bundesländern. Bundesforschungsanstalt für Forst und Holzwirtschaft. Arbeitsbericht aus den Institut für Öconomie. Aussenstelle Eberswalde 94(2). 31 p.

5.12. GREECE

In co-operation with Prof. Ioannis Meliadis

Introduction

The undulating landscape of Greece in combination with the Mediterranean climate and considerable rainfall provide very favourable conditions for forest growth. Only the summer drought limits the forest development and makes it susceptible to fires. Much of the forest and other wooded land (6.6 million ha) consists of a shrub-like, grazed vegetation (UN-ECE/FAO 2000). The species richness is very high in Greek forests and other wooded land. Some 3 million ha are seen as forest available for wood supply, and of that almost half is managed as coppice (Smiris 1999). Some 49 000 ha of forest and other wooded land burned annually between 1989 and 1998, mainly due to human causes. The annual fellings are low, and mainly consist of fuelwood. The forestry sector contributes only 0.17% to the GDP. However, the externalities from tourism (partly for forests and landscape) contribute much more.

Country specific scenario assumptions

Since only a limited dataset was available from the Nilsson et al. (1992) study, it was decided to apply a simple balance method to the UN-ECE/FAO (2000) data. The forest was initialised with a forest area of 3.09 million ha and a growing stock of 45.2 m^3/ha. The balance method is based on simple calculations of a gross annual increment of 2.5% of the growing stock when the growing stock is less than 150 m^3/ha. If the growing stock is higher, then a relative growth of 2% is used. Natural mortality is calculated in a comparable way: if the growing stock is less than 150 m^3/ha, then 7.9% of gross annual increment is assumed to be lost, otherwise the natural mortality is 10% of gross annual increment.

Business-as-usual scenario

To assume a basic fellings level, several data sources were consulted. ETTS-V foresaw that fellings would remain very stable at the 1990 level of 3.38 million m^3/yr (which was 101% of the net increment,

Pajuoja 1995). UN-ECE/FAO (2000) report annual overbark removals of 2.41 million m^3/yr. This, as the most recent source of information, was used for the fellings for the initial year. In the BAU scenario it was assumed that the fellings will stay at this level throughout the simulation period, and that the area of forest will stay at the 1990s level.

EFISCEN-ETTS scenario

ETTS-V foresaw that fellings would remain very stable at the 1990 level of 3.38 million m^3/yr (Pajuoja 1995), which was followed in this scenario.

Maximum sustainable production scenario

The fellings were increased from the initial amount to 3.24 million m^3/yr by 2005; the level under which the growing stock remained approximately stable.

Multi-functional scenario

In this scenario the UN-ECE/FAO (2000) fellings level was used as a starting point for 1990, i.e. 2.41 million m^3/yr. Then it was assumed that fellings will increase by 1% per 5 years throughout the whole simulation period. By 2050 fellings amount to 2.7 million m^3/yr. The forest area was kept at the 1990 level.

Results for Greece

Overall, the results for Greece should be regarded with care. These simple projections for 3.09 million ha of forest indicate that a maximum sustainable production of 3.24 million m^3/yr may be achieved, far higher than at present. However, the rather simple approach used here (with a relative growth rate compared to the growing stock) determines the outcome to certain extent. That is, a rather arbitrary point of 150 m^3/ha was chosen as the point where the relative growth rate is reduced and where the natural mortality increases. Nevertheless the relative growth rate was initiated in such a way that it matched the statistics of UN-ECE/FAO (2000) which reported an average increment of 1.14 m^3/ha/yr. However, Smiris (1999) reported an average increment in Greece of 2.76 m^3/ha/yr. If this latter number (which seems more realistic) had been used, it would have influenced the results considerably.

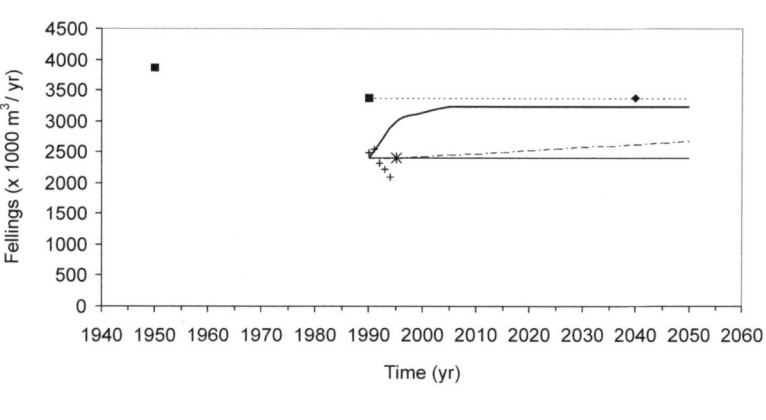

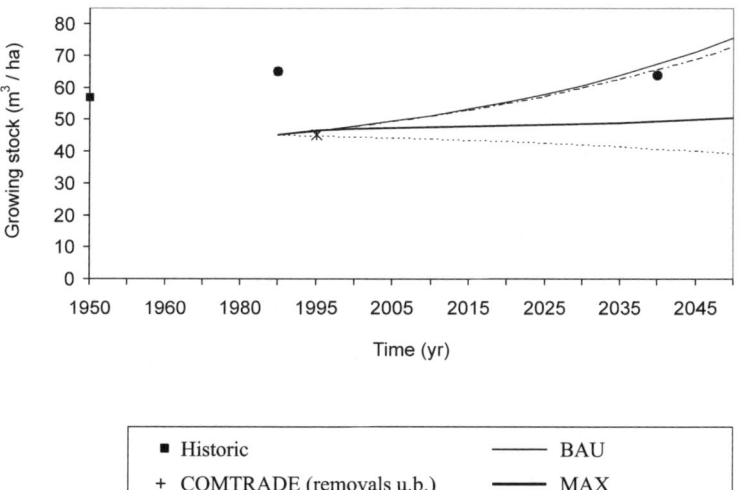

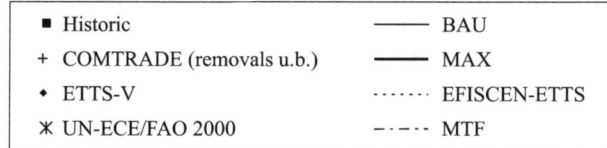

Fig. 5.33. Projected total national fellings (top) and growing stock per ha (bottom) in Greek forests under the four scenarios until 2050.

The approach of relative growth rate compared to growing stock also leads to increases in gross annual increment in those scenarios where growing stock continues to increase. By 2050 increments amount to 1.26 and 1.82 m^3/ha/yr for the MAX and the MTF scenarios, respectively.

98   COUNTRY LEVEL RESULTS

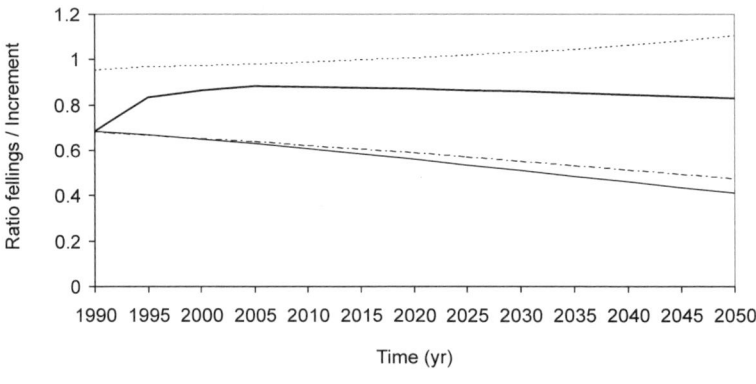

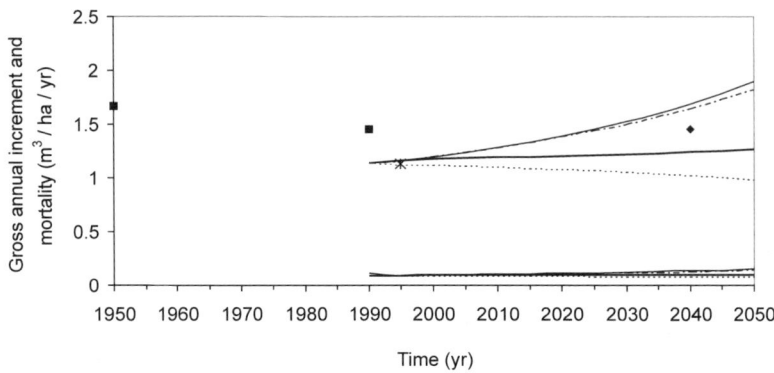

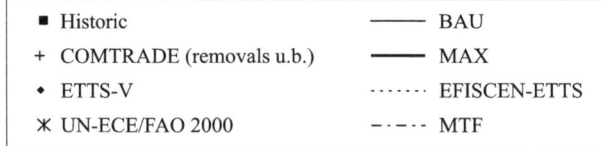

Fig. 5.34. Ratio of projected fellings/increment in Greek forests under the four scenarios until 2050 (top) and projected net annual increment and mortality (bottom).

Given the simple assumptions behind the method used here, the results indicate that Greek forests could not follow the projected fellings under the EFISCEN-ETTS scenario with required annual fellings of around 108% of the increment by 2050. The projected average growing stock may then decline to 39 m³/ha in 2050, whereas

ETTS-V projected that it may stay at their initial level of 64 m³/ha. If fellings stay at the level of the 1990s (BAU), average growing stocks may increase to 76 m³/ha. Mortality rates may then amount to 0.14 m³/ha/yr. By 2050 growing stock under the MTF scenario amounts to 73 m³/ha.

Given the low average increment that was used here, the MTF scenario show good possibilities for some recovery of the forest sector in Greece.

REFERENCES

Nilsson, S., Sallnäs, O. and Duinker, P. 1992. Future forest resources of Western and Eastern Europe. International Institute for Applied Systems Analysis. The Parthenon Publishing Group. England. 496 p.

Smiris, P. 1999. Greece. In Pelkonen, P., Pitkänen, A., Schmidt, P., Oesten, G., Piussi, P. and Rojas, E. (eds.), Forestry in Changing Societies in Europe. Part II. SILVA Network. University Press, Joensuu, Finland. Pp. 139–154.

Pajuoja, H. 1995. The outlook for the European Forest Resources and roundwood supply. Geneva, Timber and Forest Discussion Papers ECE/TIM/DP/4. UN-ECE/FAO. Geneva. 59 p.

UN-ECE/FAO 2000. Forest Resources of Europe, CIS, North America, Australia, Japan and New Zealand. Geneva, Timber and Forest Study Papers, No 17. 445 p.

Further information

Meliadis, I. 1994. New edition of the Greek forest inventory, 1960–1987. Vassilika. 272 p.

5.13. HUNGARY

In co-operation with P. Csoka

Introduction

Hungary is located in the Carpathian Basin, with hilly areas to the north and the southwest, around the central plain. It is in these hilly areas where the forests are concentrated. There, mostly rich hardwood forests occur in which different oak species and black locust (*Robinia pseudoacacia*) dominate. Of the 1.81 million ha of forest, 70% was publicly owned, and co-operatives owned 30% (UN-ECE/FAO 2000). Since the political changes, privatisation of both of these has taken place resulting in 33% of the forest in private ownership today. Most of those private owners tended to further under-utilise their wood resources (Gál and Mészáros 1999). There is a large interest in nature-oriented forest management, and the Hungarian forest policy aims to afforest about 20 000 ha/yr.

Country specific scenario assumptions

The data underlying the projections covered 1.61 million ha of forest out of the 1.7 million ha of forest available for wood supply (UN-ECE/FAO 2000). The data represent the state of the forest as of 1995. Since the areas under regeneration were not taken up in our data, all scenarios start with an afforestation in the first period to arrive at 1.7 million ha. The data were distinguished in three site classes and five tree species.

Business-as-usual scenario

To initialise the required national production, several statistics were used. UN-ECE/FAO (2000) reports a production of 6.049 million m^3 roundwood/yr, of which 5.3 million m^3/yr is from deciduous species. UN-ECE/FAO (1992) reported a production of 7.78 million m^3/yr, but that was before the political changes. ETTS-V, which was initialised also for 1990, starts with annual fellings of 7.03 million m^3/yr. The COMTRADE removals data also show the sharp decline in Hungary, from 6.1 million m^3/yr in 1990 to 3.6 million m^3/yr in 1997. The projections were thus initialised with fellings of 5.3 million

m^3/yr. Some 30% of total fellings from conifers was assumed to come from thinnings. For deciduous species 25% was assumed to come from thinnings. The area of forest was expected to stay at the 1990s level, and it was assumed that there is no change in species composition.

EFISCEN-ETTS scenario

ETTS-V foresaw a very sharp increase in fellings in Hungary, from 7 million m^3/yr in 1990 to 12 million m^3/yr in 2040. An attempt was made to follow this pattern in this scenario. A large forest area expansion of 459 000 ha between 1990 and 2040 was followed, ending with 2.065 million ha of forest in 2030.

Maximum sustainable production scenario

The fellings were quickly increased to arrive at a point where fellings equal increment; 10.9 million m^3/yr by 2015.

Multi-functional scenario

The initial annual fellings of the BAU scenario was applied, and then it was assumed that the fellings will first of all stay at the present low level until 2005. Then it was assumed that the current policy to increase wood production will work, and fellings rise sharply, with 28% per 5 years between 2005 and 2010. After that the increase slows down to 10% per 5 years. After 2020, the rise slows down further to 6% per 5 years until 2040 to arrive at a level of 10.8 million m^3/yr. After that fellings stabilise. To take into account the trend towards nature-oriented forest management, the rotation lengths were elongated by 20 years. The share of thinnings from total fellings gradually increased to 40%. The area initially over 80 years of oak and other hardwood deciduous species, was set aside. Initially this was 245 000 ha. Through ageing this can increase during the simulation. Afforestation was assumed to take place at half the pace that ETTS-V foresaw.

Results for Hungary

The projections for Hungary for the 1.7 million ha of forests show that a maximum sustainable production of 10.9 million m^3/yr may be achieved, under which the present growing stock of 192 m^3/ha increases to 198 m^3/ha by 2050. Given the large interest in Hungarian forest policy to increase forest area and production,

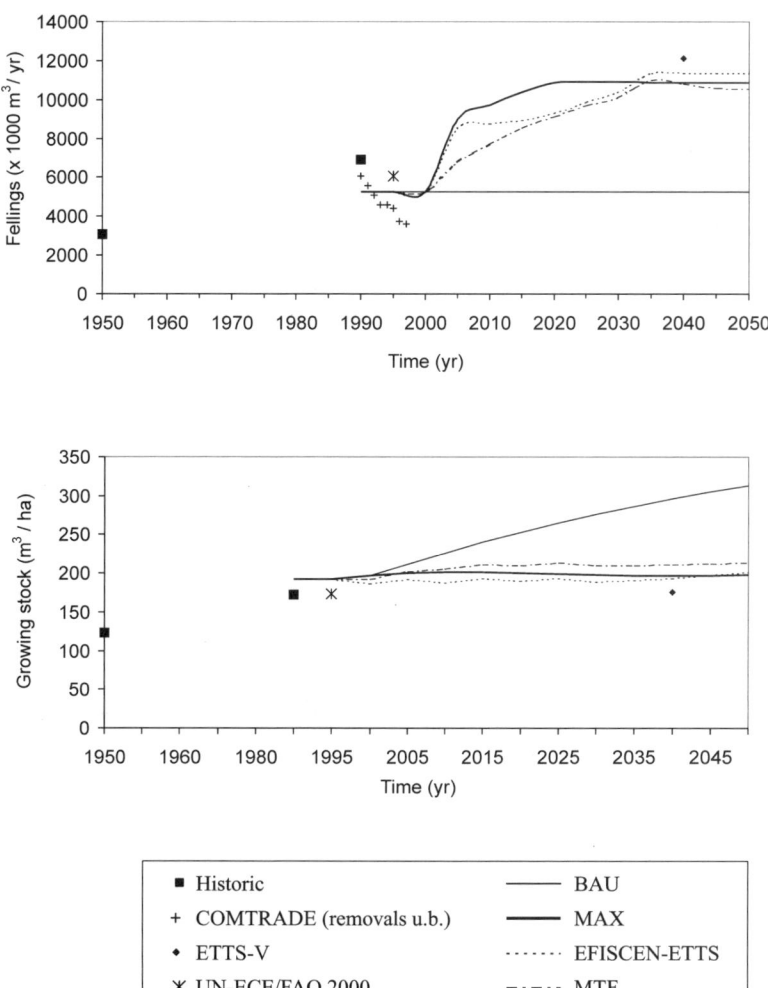

Fig. 5.35. Projected total national fellings (top) and growing stock per ha (bottom) in Hungarian forests under the four scenarios until 2050.

both the EFISCEN-ETTS and the MTF reflect that desire. The combination of a more nature-oriented forest management with a scenario where the economy will recover quickly, with large increases in fellings as a result, was reflected in the MTF scenario. Under both the EFISCEN-ETTS and the MTF scenarios, fellings reach the biological maximum by 2035.

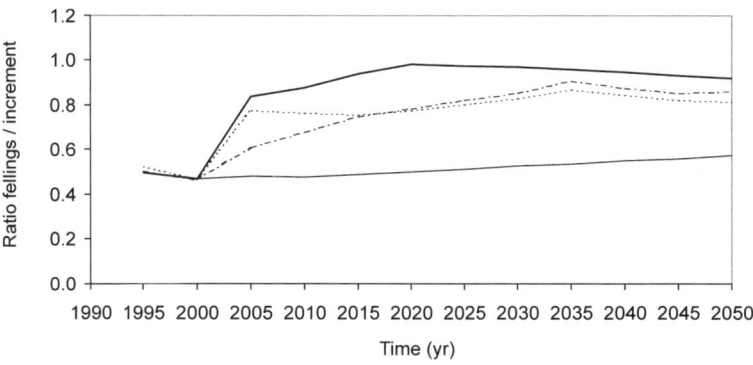
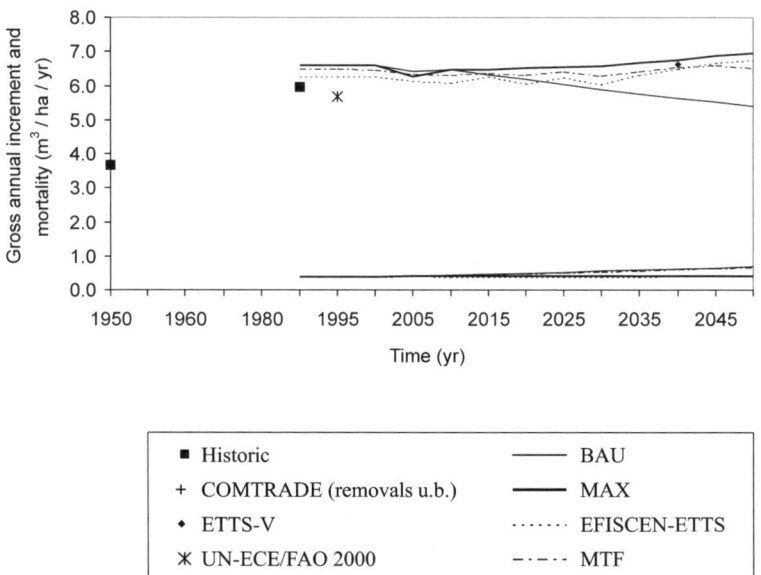

Fig. 5.36. Ratio of projected fellings/increment in Hungarian forests under the four scenarios until 2050 (top) and projected net annual increment and mortality (bottom).

The growing stock does not increase much under these two scenarios, and may amount to 202 and 214 m³/ha by 2050 for the EFISCEN-ETTS and the MTF scenarios, respectively. Under the BAU it may amount to 313 m³/ha, and mortality rates to 12.6% of the gross annual increment. The gross annual increment declines only

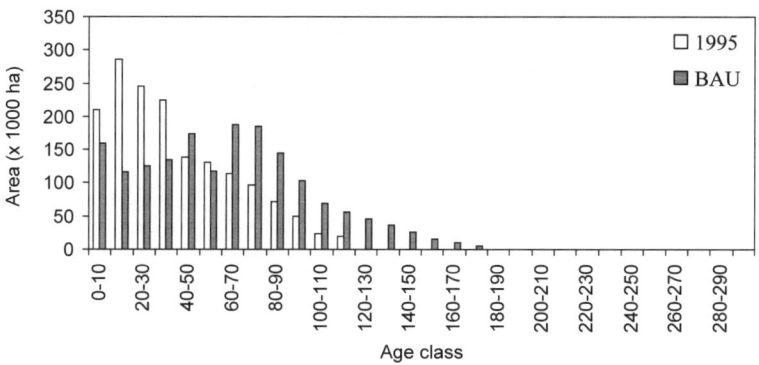

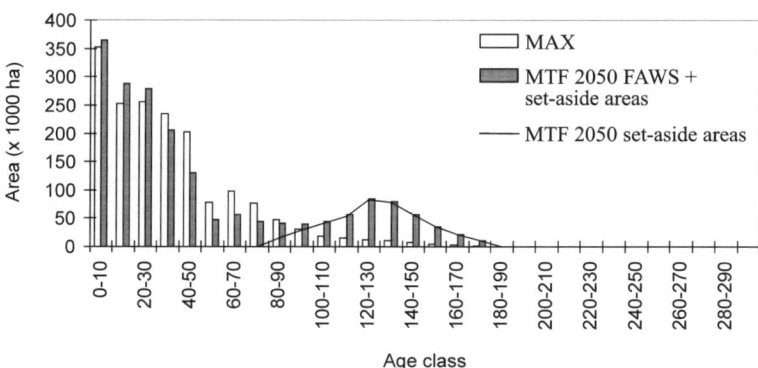

Fig. 5.37. The age class distribution of the Forest Available for Wood Supply (FAWS) in the initial situation and for 2050 under the BAU scenario (top). The age class distribution of FAWS for 2050 under the MAX scenario, the MTF scenario including set-aside areas and the MTF scenario with only set-aside areas (bottom).

under the BAU scenario from 6.6 m³/ha/yr initially, to 5.4 m³/ha/yr in 2050; this is because the build-up of growing stock is the largest here. Under all other scenarios the gross annual increment is very stable or increases slightly during the simulated period.

The age class distribution for 2050 clearly shows the ageing process under the BAU scenario because the utilisation ratio is around 55% only. The ageing process does not occur under the MTF and the MAX scenarios. In the MTF scenario 424 000 ha (22% of total forest area) are projected to be newly established as set-aside areas by 2050. Still, the large increase in fellings as foreseen for the MTF scenario was possible.

REFERENCES

Gál, J.K. and Mészáros, J. 1999. Hungary. In: Pelkonen, P., Pitkänen, A., Schmidt, P., Oesten, G., Piussi, P. and Rojas, E. (eds.), Forestry in Changing Societies in Europe. Part II. SILVA Network. University Press, Joensuu, Finland. Pp. 155–165.

UN-ECE/FAO 1992. The Forest Resources of the Temperate Zones, the UN-ECE/FAO 1990 Forest Resource Assessment, General Forest Information. Geneva, Switzerland. 347 p.

UN-ECE/FAO 2000. Forest Resources of Europe, CIS, North America, Australia, Japan and New Zealand. Geneva, Timber and Forest Study Papers, No 17. 445 p.

5.14. IRELAND

In co-operation with Mr. Pat Farrington

Introduction

The Atlantic climate provides Ireland with plenty of rainfall, and thus in principle good growing conditions. Many of the forest sites consist of gleyic and peaty soils. The 591 000 ha of forest is mostly publicly owned, but afforestations are now mostly done by private land owners, resulting in a changing ownership structure. Most of the forest consists of plantations of Sitka spruce that is managed in rather short rotation periods of 40–45 years (Nieuwenhuis 1999). This is, of course, also reflected in the fellings that consist of 98% of coniferous wood. The forestry sector contributes only 0.3% to the GDP. Ireland has a forest policy that aims at further expansion of the forest area, investments in forest industry, but also at a wider diversity of tree species, and a more nature-oriented forest management.

Country specific scenario assumptions

UN-ECE/FAO (2000) presents a total forest area of 591 000 ha in Ireland. The data underlying the projections here for Ireland only include the coniferous forests that are under supervision of the private limited company Coillte which manages the public forests; 344 000 ha, which is 92% of the total public forest. This is also very close to the area that ETTS-V used. The data represent the state of the forest as of 1992–1993. The data were distinguished by five site classes and seven coniferous tree species.

Business-as-usual scenario

Several statistics were used to initialise fellings at the present level. UN-ECE/FAO (2000) reports a production of 2.3 million m^3 roundwood/yr, of which 2.29 million m^3/yr is from coniferous species. UN-ECE/FAO (1992) reported a production of 1.4 million m^3/yr. ETTS-V, which was initialised also for 1990, starts with annual fellings of 1.44 million m^3/yr. The COMTRADE removals data show a steady increase in Ireland, from 1.6 million m^3/yr in 1990 to 2.2 million m³/yr in 1997. Since we do not cover the private forests in the underlying data, the projections were initialised with 85% of the fellings according to UN-ECE/FAO (2000): 1.92 million m³/yr. Some 24% of total fellings from conifers was assumed to come from thinnings. It was assumed that the area of forest will stay at the 1990s level and that there is no change in species composition.

EFISCEN-ETTS scenario

ETTS-V foresaw a very sharp increase in fellings in Ireland, from 1.44 million m³/yr in 1990 to 6.1 million m³/yr in 2040. An attempt was made to follow this for the EFISCEN-ETTS scenario, but with a levelling off in 2040. An area correction was also carried out to take into account real afforestations (amounting to 153 000 ha) between 1992 and 1999. ETTS-V also foresaw a large forest area expansion of 310 000 ha between 1990 and 2040; that was also followed for this scenario. Two-thirds of these afforestations are carried out with Sitka spruce, and the rest with Scots pine, ending up with 0.73 million ha of forest in 2030 (starting from the coniferous forests of Coillte only).

Maximum sustainable production scenario

The fellings (on the initial area of 344 000 ha) were quickly increased to arrive at a point where fellings equal increment; i.e. a level of 4.4 million m³/yr by 2005.

Multi-functional scenario

The initial annual fellings of the BAU scenario was applied. An area correction was also carried out to take into account real afforestations (amounting to 153 000 ha) between 1992 and 1999. Then it was assumed that the fellings will follow the ETTS-V scenario, but level off earlier (around 2015) to reach a level of 3.8 million m³/yr. Since the real afforestations in the 1990s were around 17 000 ha/yr, it was assumed that the ETTS-V forest expansion scenario is close to feasible. Therefore, it was reduced slightly, ending with 0.65 million ha of forest in 2035 (starting from the coniferous forests of Coillte only).

To take into account the desire for a more nature-oriented forest management, the rotation lengths were prolonged by 10 years. The share of thinnings from total fellings gradually increased to 40%, and afforestations are now done for more than half with Scots pine. The latter should represent nature-oriented forestry. We set aside Scots pine and lodgepole pine over 50 years of age.

Since there were no broadleaved species in the data, it is not possible to set aside any area in those or afforest with those species, although the forestry action plan indicates that. Initially 3000 ha was set aside. Through ageing this can increase during the simulation.

Results for Ireland

When looking at the Irish results one has to keep in mind that in both the EFISCEN-ETTS and MTF scenarios a large-scale afforestation was carried out. This was not the case for MAX and the BAU scenarios. Therefore, the projections for Ireland show that a maximum sustainable production of 3.3 million m^3/yr may be achieved (on present area), lower than that which is finally achieved under the EFISCEN-ETTS and the MTF scenarios. Therefore, the results show that the impact of relatively large afforestations in rather short rotations can be seen already within this time frame of projections of 60 years. The MAX also shows that felling levels are sensitive to ageing when working with these rather short rotations and short age class intervals where fellings can be carried out.

Given the large interest in Irish forest policy to increase forest area and production, both the EFISCEN-ETTS and the MTF reflect that strategy. Under the EFISCEN-ETTS and the MTF, the fellings stabilise at 4.6 and 3.8 million m^3/yr, respectively. Gallagher and O'Carroll (2001) foresaw that the production increase will occur faster, and will increase to 5.0 million m^3/yr by 2015 already.

The large-scale afforestations in EFISCEN-ETTS and MTF also reflect the impact that such measures have on growing stock and increment. Namely afforestations add rather large areas with no growing stock and low increment, and thus reduce national averages. This can clearly be seen in the graphs for growing stock and increment.

Therefore, the growing stock does not increase much under EFISCEN-ETTS and the MTF scenarios: respectively, 202 and 246 m^3/ha by 2050. Under the BAU it may amount to 421 m^3/ha. Under this latter scenario, mortality rates may then amount to 7% of the gross annual increment; producing 0.2 million m^3 of dead wood per year. The gross annual increment first increases a lot under the BAU scenario from 9.4 m^3/ha/yr initially, to 12.9 m^3/ha/yr in 2025,

COUNTRY LEVEL RESULTS

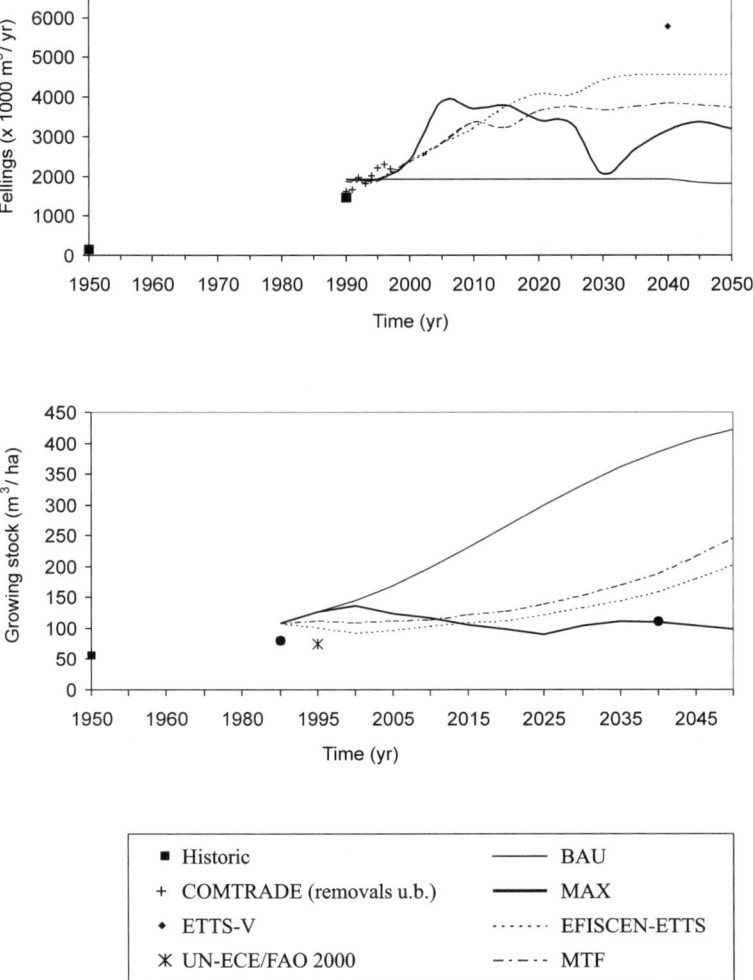

Fig. 5.38. Projected total national fellings (top) and growing stock per ha (bottom) in Irish forests under the four scenarios until 2050.

and then declines again to 8.9 m^3/ha/yr in 2050. This is clearly an effect of ageing.

The age class distribution for 2050 clearly shows the ageing process under the BAU scenario because the utilisation ratio is around 50% only. The ageing process occurs under the MTF scenario as well, but

IRELAND

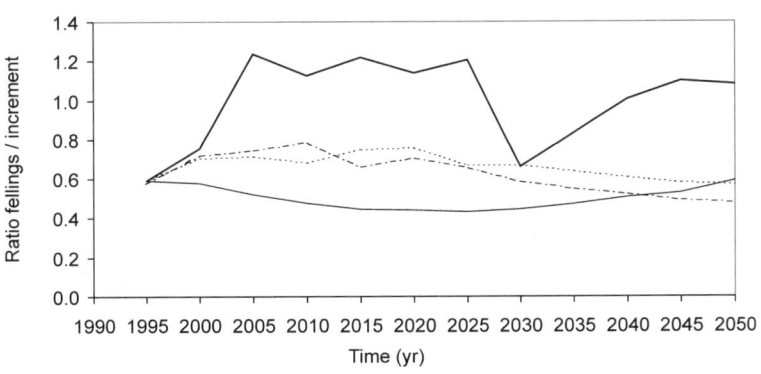

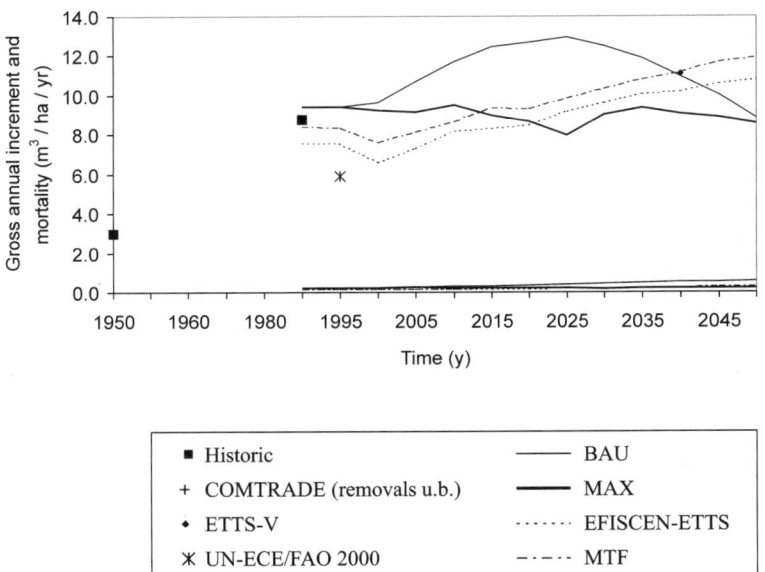

Fig. 5.39. Ratio of projected fellings/increment in Irish forests under the four scenarios until 2050 (top) and projected net annual increment and mortality (bottom).

is not very visible because there was a large area of new afforestation. In the MTF scenario 27 000 ha are projected to be newly established as set-aside areas by 2050. Still, the increase in fellings as foreseen for the MTF scenario was possible.

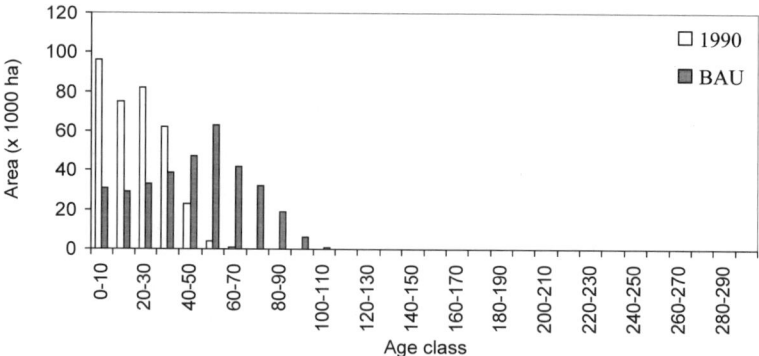

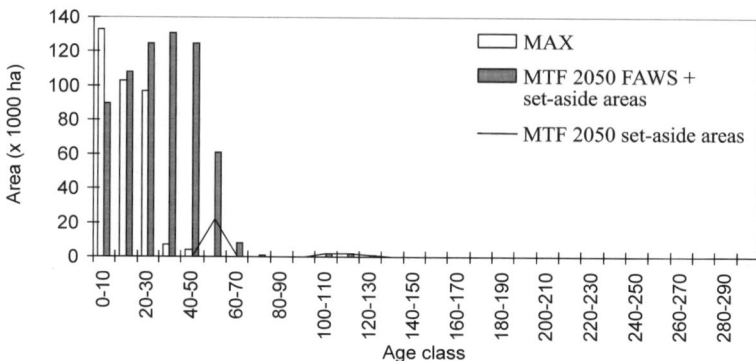

Fig. 5.40. The age class distribution of the Forest Available for Wood Supply (FAWS) in the initial situation and for 2050 under the BAU scenario (top). The age class distribution of FAWS for 2050 under the MAX scenario, the MTF scenario including set-aside areas and the MTF scenario with only set-aside areas (bottom).

REFERENCES

Gallagher, G. and O'Carroll, J. 2001. Forecasts of roundwood production from the forests of Ireland 2001–2015. COFORD. 24 p.

Nieuwenhuis, M. 1999. The Republic of Ireland. In: Pelkonen, P., Pitkänen, A., Schmidt, P., Oesten, G., Piussi, P. and Rojas, E. (eds.), Forestry in Changing Societies in Europe. Part II. SILVA Network. University Press, Joensuu, Finland. Pp. 167–186.

UN-ECE/FAO 1992. The Forest Resources of the Temperate Zones, the UN-ECE/FAO 1990 Forest Resource Assessment, General Forest Information. Geneva, Switzerland. 347 p.

UN-ECE/FAO 2000. Forest Resources of Europe, CIS, North America, Australia, Japan and New Zealand. Geneva, Timber and Forest Study Papers, No 17. 445 p.

Further information

Department of Agriculture, Food, and Forestry. 1996. Growing for the future, A strategic plan for the development of the forestry sector in Ireland. Dublin, Ireland. 98 p.

O'Carroll, J. and Gallagher, G. 2001. Forecast of Roundwood Production from the Forests of Ireland 2001–2015. COFORD, Ireland. 28 p.

5.15. ITALY

In co-operation with Dr. Franco Cozza

Introduction

Two mountain ranges shape the Italian territory: the Alps in the north and the Apennines stretching north-south. It is these mountain ranges where the forests are concentrated. This in combination with the Mediterranean climate provides good growing conditions for forests. UN-ECE/FAO (2000) report a present forest area of 9.86 milllion ha, of which two-thirds are privately owned. Amongst the private owners, coppice is a widespread management system (Colpi et al. 1999). However, more and more of the coppice forests are left to outgrow to high forest. The forest available for wood supply is 6 million ha. In the Alps, selective management systems are common. Italian forests are amongst the most diversified and richest semi-natural forests in Europe. However, the interests of forest owners are also extremely diverse, and this has led to a situation where Italy has both a large and rapidly increasing growing stock, but where it is also one of the larger net importers of wood products in Europe.

Country specific scenario assumptions

The data underlying the projections covered 3.8 million ha of forest by age class distributions and 1.9 million ha by diameter class distribution. Out of the 3.8 million ha of even-aged forest, 2.86 million ha was under coppice type of management. Thus the two modelling approaches were used and the results were aggregated afterwards. The data represent 85% of the forest area according to UN-ECE/FAO (1992). The data are based on the 1985 inventory cycle. The age class data were distinguished by 19 tree species. The diameter class data were distinguished by three site classes and ten tree species.

Business-as-usual scenario

UN-ECE/FAO (2000) reports an annual felling of 8.75 million m^3 roundwood/yr, of which 6.9 million m^3/yr is from deciduous species. The COMTRADE removals data give a slightly increasing trend from 8.0 million m^3/yr in 1990 to 9.2 million m^3/yr in 1997. ETTS-V, which was initialised also for 1990, starts with annual fellings of 8.6 million m³/yr. Taking into account the share of forest covered in the present simulations, the fellings were initialised with 5.5 million m³/yr from the even-aged part of the forest, and 0.98 million m³/yr from the uneven-aged part. In the even-aged high forest, 30% of total fellings was assumed to come from thinnings. For coppice 20% was assumed to come from thinnings. It was assumed that the area of forest and its species distribution will stay at the 1990s level.

EFISCEN-ETTS scenario

ETTS-V foresaw a gradual decrease in fellings in Italy until 2010, and a slight increase afterwards; ending at the same level as in 1990: i.e. 8.6 million m³/yr. This pattern was followed for this scenario. ETTS-V also foresaw an area decrease in exploitable forest of 90 000 ha by 2000, and then a gradual increase of 100 000 ha over the period until 2040. It was therefore assumed that the area of forest will stay at the 1990s level.

Maximum sustainable production scenario

The fellings were quickly increased to arrive at a point where fellings equal increment; i.e. a level of 16.2 million m³/yr by 2010.

Multi-functional scenario

The initial annual fellings of the BAU scenario was applied, and then an increase in fellings of 1% per year until 2020 was assumed. After that it stabilises. This is based on the assumption that Italian forest policy will succeed in improving the national self-sufficiency. The increase was assumed for the even-aged part of the forest, but equally for high forest and coppice. To emulate current discussion concerning nature-oriented management, the rotation lengths were prolonged by 20 years in high forest and by 10 years in coppice (to assume outgrowth of coppice). The share of thinnings from total fellings gradually increased to 40% in high forests, and 30% from deciduous coppice. The area initially over 140 years of high forest of 'other oak' and 'other broadleaved', and areas over 70 years of coppice of turkey oak (*Quercus cerris*) and holly oak (*Q. ilex*) were set aside. Through ageing this can increase during the simulation. An afforestation of 180 000 was assumed to take place before 2010 in high forest, equally divided between spruce and beech.

Results for Italy

Most apparent from the Italian results is that the current fellings are far below the maximum sustainable level of 19.4 million m³/yr. Under this level the growing stock slightly increases to 154 m³/ha in 2050, and natural mortality amounts to rates of 8.7% of the gross annual

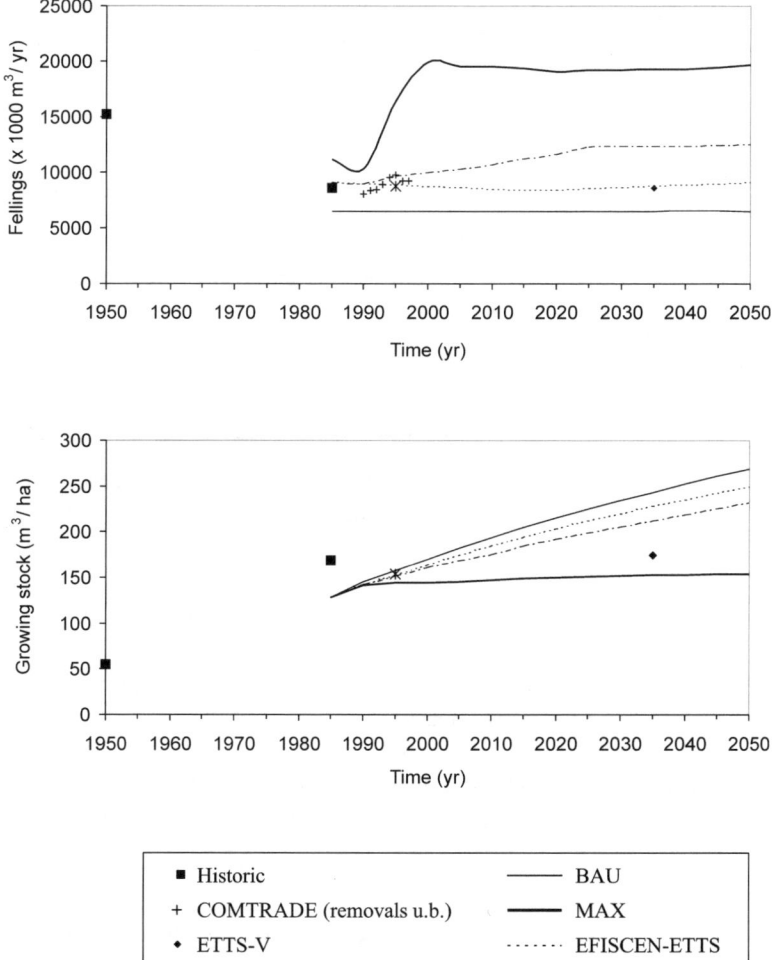

Fig. 5.41. Projected total national fellings (top) and growing stock per ha (bottom) in Italian forests under the four scenarios until 2050.

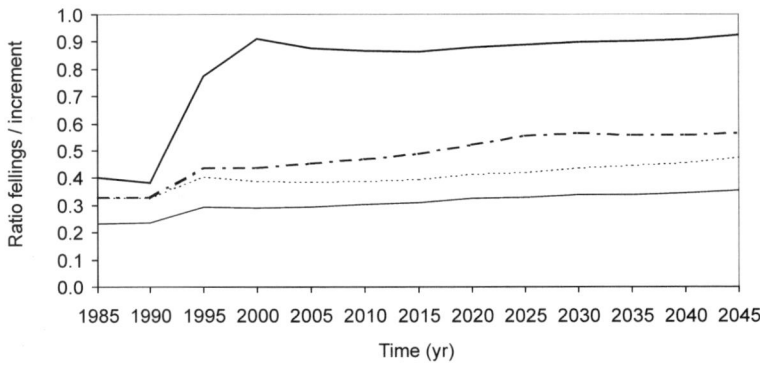

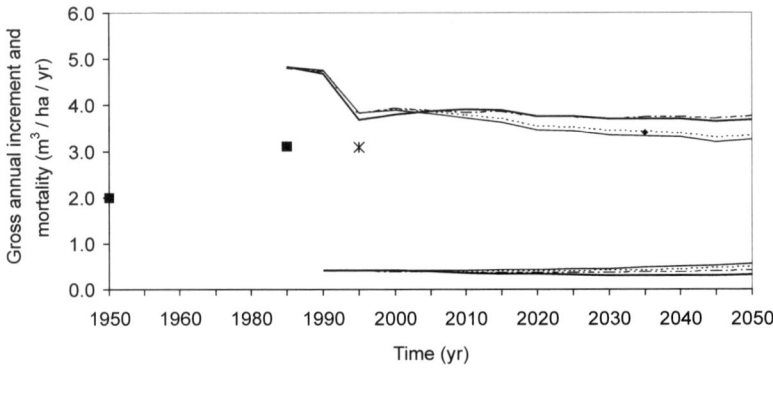

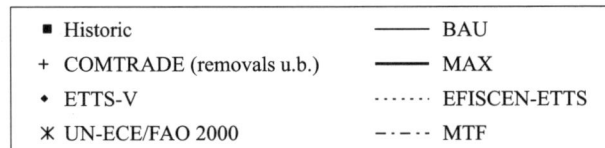

Fig. 5.42. Ratio of projected fellings/increment in Italian forests under the four scenarios until 2050 (top) and projected net annual increment and mortality (bottom).

increment. Under this scenario a much more evenly distributed age class distribution is reached by 2050.

Growing stocks rise sharply to 232 m³/ha and 269 m³/ha for the MTF and the BAU scenarios, respectively. A mortality rate of 12% of the increment, or a total production of dead wood of 2.5 million

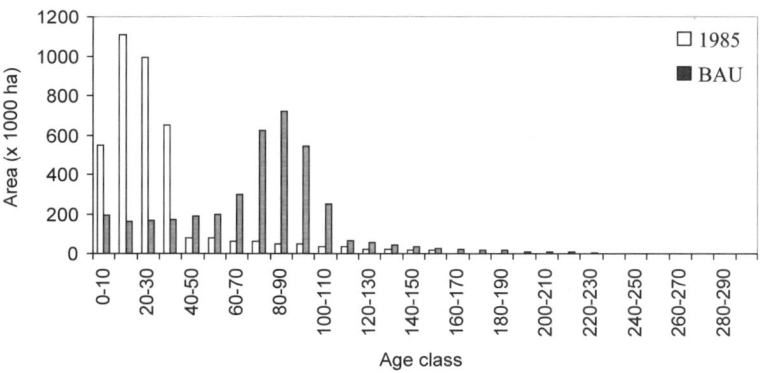

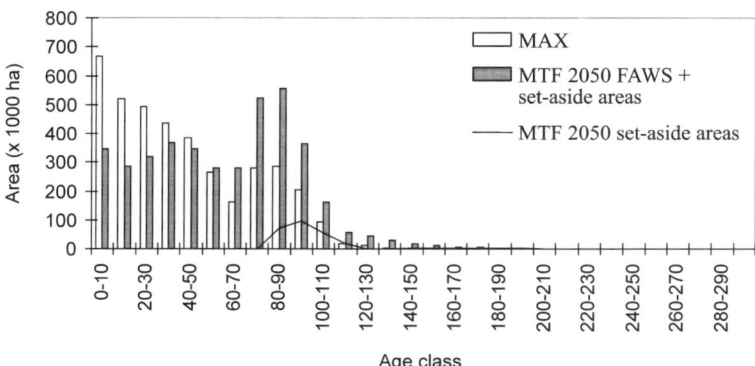

Fig. 5.43. The age class distribution of the Forest Available for Wood Supply (FAWS) in the initial situation and for 2050 under the BAU scenario (top). The age class distribution of FAWS for 2050 under the MAX scenario, the MTF scenario including set-aside areas and the MTF scenario with only set-aside areas (bottom).

m^3/yr on the 5.7 million ha of simulated forest may then be achieved under the MTF scenario.

The gross annual increment always shows a drop at the beginning of each scenario, and is rather stable afterwards at around 3.5 m^3/ha/yr. The sudden drop is caused by rather large concentrations of areas that move to a next age class from one time period to the next. Still, this NAI is higher than the latest statistic reported for Italian increment to UN-ECE: 3.1 m^3/ha/yr.

In the age class distributions for 2050, all scenarios (except for the MAX scenario) show the outgrowth of the concentration of area of coppice. For the MAX scenario a more evenly distributed area over age classes is generated. The MTF scenario also shows the established reserves; an area of 257 000 ha is projected to be set aside in the even-aged part of the forest by 2050.

REFERENCES

Colpi, C., Pettenella, D., Urbinati, C. and Cavalli, R. 1999. Italy. In: Pelkonen, P., Pitkänen, A., Schmidt, P., Oesten, G., Piussi, P. and Rojas, E. (eds.), Forestry in Changing Societies in Europe. Part II. SILVA Network. University Press, Joensuu, Finland. Pp. 187–201.
UN-ECE/FAO 1992. The Forest Resources of the Temperate Zones, the UN-ECE/FAO 1990 Forest Resource Assessment, General Forest Information. Geneva, Switzerland. 347 p.
UN-ECE/FAO 2000. Forest Resources of Europe, CIS, North America, Australia, Japan and New Zealand. Geneva, Timber and Forest Study Papers, No 17. 445 p.

Further information

Ministerio dell'Agricoltura e delle Foreste, 1988. Inventario Forestale Nazionale 1985. Sintesi metodologica e risultati. I.S.A.F.A. 461 p.

5.16. LUXEMBOURG

In co-operation with Mr. Mark Wagner

Introduction

Although the Luxembourg forests comprise only 0.04% of the European forests, nationally they are of great importance. The 85 000 ha covers 33% of the land, and the forests are important for scenic amenity, recreation and tourism in the hilly landscape of Luxembourg. The Atlantic climate provides good growing conditions, and together with a long tradition of forest management this has created an average growing stock of 240 m^3/ha and an increment of 7.9 m^3/ha/yr (UN-ECE/FAO 2000). The forests are equally divided between public and private ownership.

Country specific scenario assumptions

UN-ECE/FAO (2000) presents a total forest area of 86 000 ha in Luxembourg. The data underlying the projections here cover 71 000 ha only and do not include the coppice forests. The data represent the state of the forest as of 1989. The data were distinguished by three deciduous tree species, and three coniferous species.

Business-as-usual scenario

Several statistics were used to initialise fellings at the present level. UN-ECE/FAO (2000) reports a removal overbark 360 000 m^3 roundwood per year, of which 180 000 m^3/yr is from coniferous species. ETTS-V, which was initialised also for 1990, starts with annual fellings of 360 000 m^3/yr. The COMTRADE removals data show a steady decrease in removals underbark in Luxembourg from 550 000 m^3/yr in 1990 to 391 000 m^3/yr in 1997. Since we do not cover the coppice forests in the underlying data, the projections were initialised with 84% of the fellings according to UN-ECE/FAO (2000): 320 000 m^3/yr. Half of these total fellings was assumed to come from coniferous forests and the share of thinnings from total fellings was set at 30%. It was assumed that the area of forest and its species distribution will stay at the 1990s level.

EFISCEN-ETTS scenario

ETTS-V foresaw a very stable fellings level in Luxembourg of 360 000 m³/yr. This was followed here, taking into account that we cover only 84% of the forest. It was assumed that the area of forest will stay at the 1990s level and that there is no change in species composition.

Maximum sustainable production scenario

The fellings were quickly increased to arrive at a point where fellings equal increment; i.e. a level of 680 000 m³/yr by 2000. The relative increase was the same for both coniferous and deciduous species.

Multi-functional scenario

The initial annual fellings of the BAU scenario was applied, and then it was assumed that the coniferous fellings will increase by 3% per year until 2005, and then level off. Deciduous fellings were assumed to stay at the 1990s level. It was assumed that some forest area expansion will be carried out with oak and beech, totalling 10 000 ha in the period 1990–2015. To take into account the desire for a more nature-oriented forest management, the rotation lengths were prolonged by 20 years. The share of thinnings from total fellings gradually increased to 40%. Oak and beech over 140 years of age were set aside; initially this is an area of 8000 ha. This can increase during the simulation as a result of ageing.

Results for Luxembourg

The projections for Luxembourg show that a maximum sustainable production of 667 000 m³/yr may be achieved, more than double the present volume of fellings. Under that high felling level, the growing stock increases slightly to 355 m³/ha by 2050, from the present 321 m³/ha.

The EFISCEN-ETTS scenario is comparable to the BAU where fellings stay at a level 320 000 m³/yr. Growing stock then rises to 572 m³/ha; the highest in Europe. Under that level of stocking, the mortality amounts to 21% of the gross annual increment; producing 1.6 million m³ of dead wood per year.

The gross annual increment declines in all scenarios due to high growing stocks being reached and the ongoing ageing. Gross

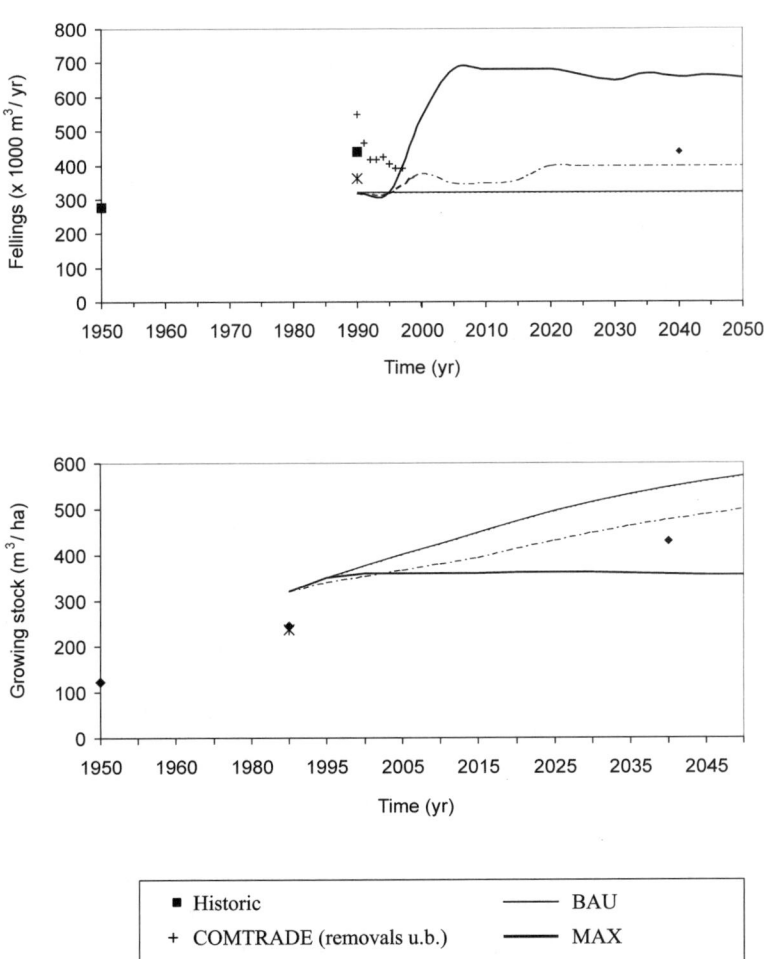

Fig. 5.44. Projected total national fellings (top) and growing stock per ha (bottom) in Luxembourgian forests under the four scenarios until 2050.

annual increment decreases from 11.6 m³/ha/yr initially, to around 9 m³/ha/yr in 2025. The decline is the least in the MAX scenario, where the ageing is the least.

The age class distributions for 2050 clearly show the ageing process under the BAU and MTF scenarios, starting already from an age class

124 COUNTRY LEVEL RESULTS

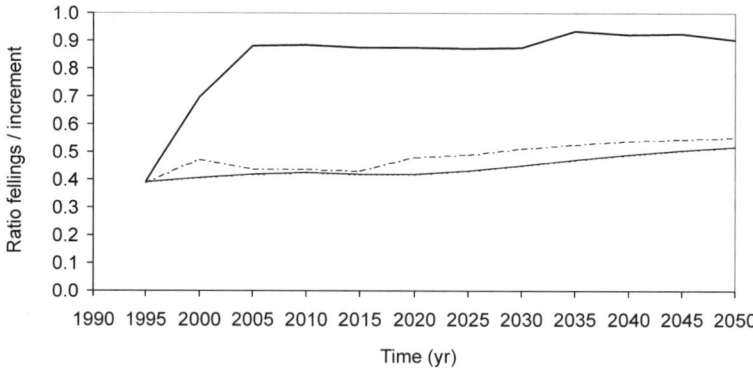

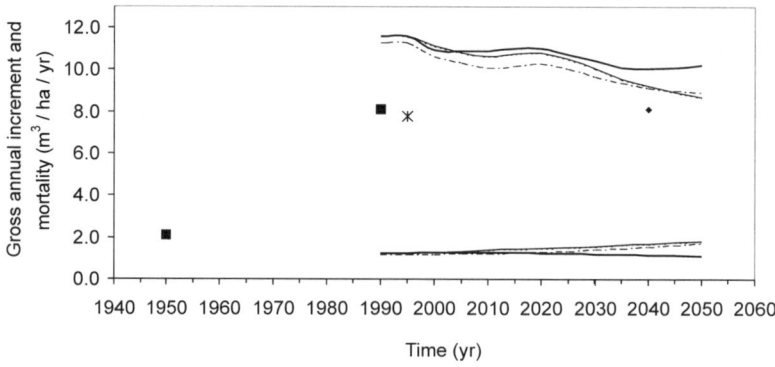

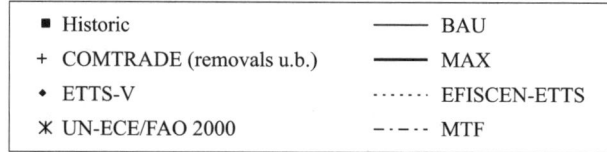

Fig. 5.45. Ratio of projected fellings/increment in Luxembourgian forests under the four scenarios until 2050 (top) and projected net annual increment and mortality (bottom).

distribution with a large proportion of old forests in 1990. In the MTF scenario, 22 000 ha are projected to be newly established set-aside areas by 2050. Still, the increase in fellings as foreseen for the MTF scenario was possible.

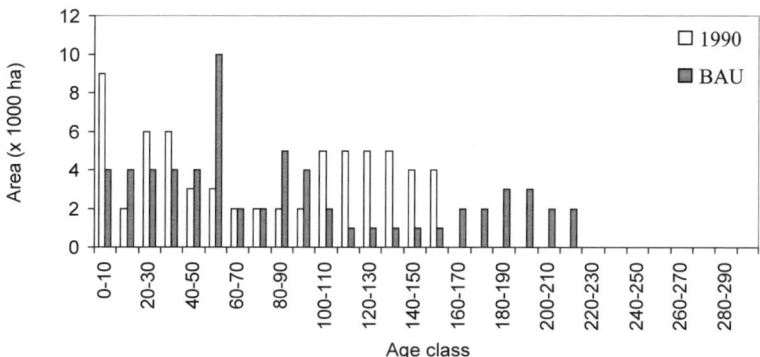

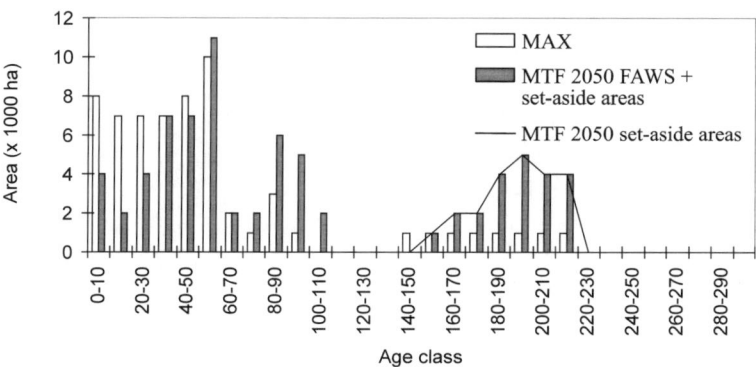

Fig. 5.46. The age class distribution of the Forest Available for Wood Supply (FAWS) in the initial situation and for 2050 under the BAU scenario (top). The age class distribution of FAWS for 2050 under the MAX scenario, the MTF scenario including set-aside areas and the MTF scenario with only set-aside areas (bottom).

REFERENCES

UN-ECE/FAO 2000. Forest Resources of Europe, CIS, North America, Australia, Japan and New Zealand. Geneva, Timber and Forest Study Papers, No 17. 445 p.

5.17. MACEDONIA

Introduction

Much of Macedonia is formed by a plateau. The plateau is surrounded by foothills and the more mountainous regions of the Dinaric Alps. Due to the large climate differentiation, the vegetation is very diverse, with many endemic plant species. Most of the 906 000 ha of forest is state owned (UN-ECE/FAO 2000), and 745 000 ha are regarded as available for wood supply. Although the forests are not very important in terms of total volume of commercial wood supply, they are very important locally for construction wood, fuelwood collection, and collection of a wide variety of non-wood products.

Country specific scenario assumptions

The data underlying the projections here for Macedonia cover 805 000 ha, which is almost 90% of the total forest. Some 85% of this area was under a coppice type management system. No new dataset was gathered, and thus the data are the same as that used by Nilsson et al. (1992). The data represent the situation around 1985 and were distinguished by two owner classes and four tree species.

Business-as-usual scenario

The information base to initialise fellings for Macedonia was rather small and uncertain due to unrecorded fuelwood fellings. UN-ECE/FAO (2000) reports a felling of 0.99 million m^3 roundwood per year, but also reports removals of 1.1 million m^3/yr. The COMTRADE removals underbark data show a decrease in Macedonia from 0.86 million m^3/yr in 1993 to 0.77 million m^3/yr in 1997. The projections were initialised with fellings according to UN-ECE/FAO (2000): 0.99 million m^3/yr, of which 85% is from coppice type management systems. For the high forests, 30% of total fellings was assumed to come from thinnings, and for coppice this was 20%. It was assumed that the area of forest will stay at the 1990s level and that there is no change in species composition.

EFISCEN-ETTS scenario

Here the same trend as for the former Yugoslavia was assumed. ETTS-V predicted a very sharp increase in fellings in Yugoslavia which would mean relative to the proportion of forest area in Macedonia, an increase from 0.99 million m^3/yr in 1990 to 2.3 million m^3/yr in 2020. An attempt was made to follow this pattern for this scenario. It was assumed that the area of forest will stay at the 1990s level.

Maximum sustainable production scenario

The fellings were quickly increased to arrive at a point where fellings equal increment; i.e. a level of 1.6 million m^3/yr by 1995 (the initial year was 1985).

Multi-functional scenario

The initial annual fellings of the BAU scenario was applied for the initial year. For the period after that it was assumed that the economic recovery of Macedonia will be very slow; commercial fellings will fall, and non-commercial fellings will stay at the level of the 1990s. Until 2005 fellings will decrease by 2% per year, and after that increase again by 2% per year until 2035. Total fellings will thus end at 1.19 million m^3/yr. A forest area decrease of 40 000 ha is also assumed in the period between 1985 and 2005. Despite the slow economic recovery, it was assumed that there will be a considerable amount of international attention focussed on Macedonian endemic species richness, and that a strong desire for nature-oriented forest management (aiming at ecotourism) will be implemented. The rotation lengths were prolonged by 20 years in high forest and by 10 years in coppice. The share of thinnings from total fellings gradually increased to 40%. We set aside broadleaved high forests over 100 years of age, and coppice over 50 years of age. Initially 75 000 ha was set aside. Through ageing this can increase during the simulation.

Results for Macedonia

When looking at the Macedonian results one has to keep in mind that the underlying database is rather old, and may no longer accurately reflect the current status of forests. It is also clearly visible from the increment graph, that estimates for increment vary considerably, thus indicating a large degree of uncertainty.

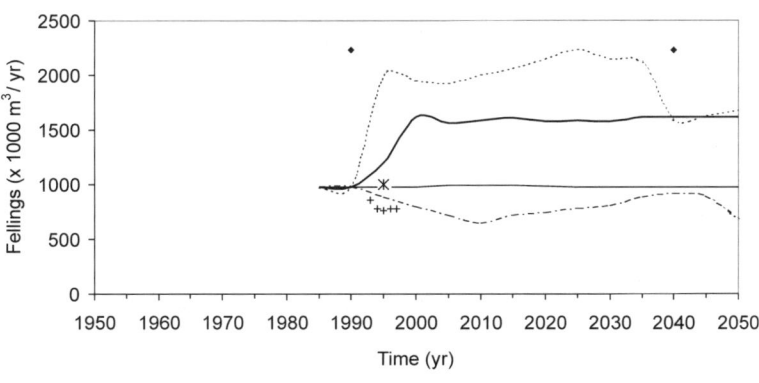

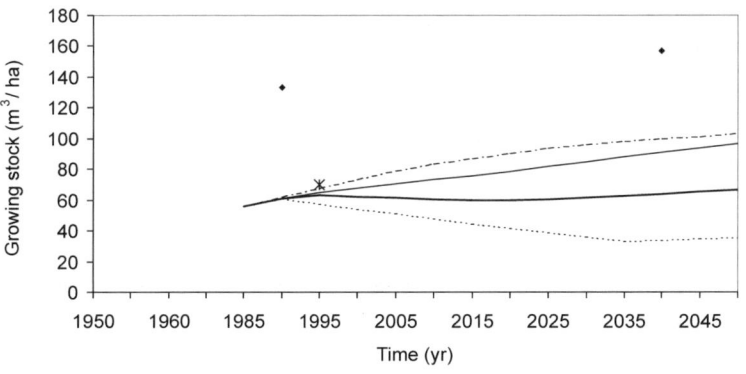

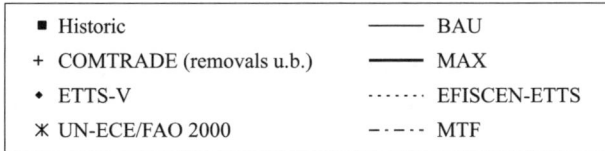

Fig. 5.47. Projected total national fellings (top) and growing stock per ha (bottom) in Macedonian forests under the four scenarios until 2050.

The projections for Macedonia show that a maximum sustainable production of 1.6 million m^3/yr may be achieved. This is much higher than the official statistics indicate for present levels. The maximum sustainable felling level is lower than the 2.3 million m^3/yr indicated by ETTS-V (EFISCEN-ETTS scenario). However, they assumed a higher gross annual increment of 3.4 m^3/ha/yr, while the data

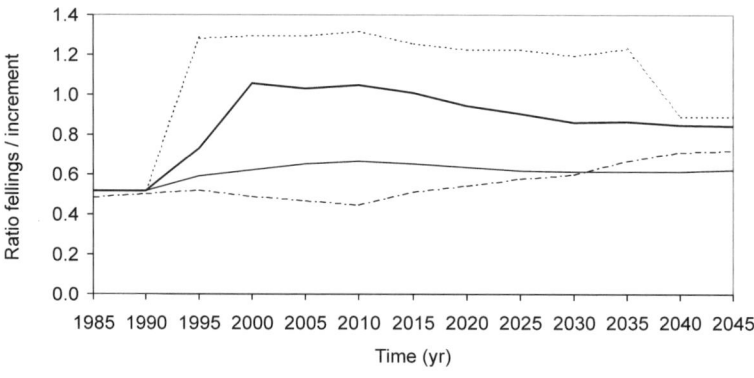

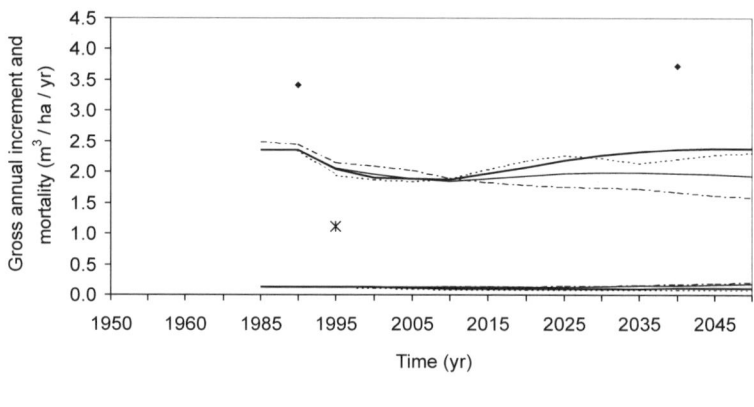

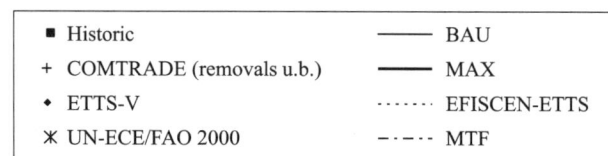

Fig. 5.48. Ratio of projected fellings/increment in Macedonian forests under the four scenarios until 2050 (top) and projected net annual increment and mortality (bottom).

underlying our projections indicate an increment of 2.4 m³/ha/yr. It remains uncertain what would be an appropriate increment level. Our underlying data indicated an increment of 2.3 m³/ha/yr for the neighbouring country Albania, providing some support for the simulated increment for Macedonia.

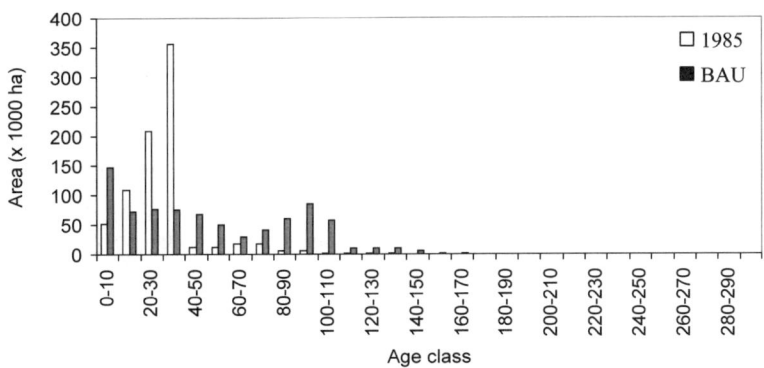

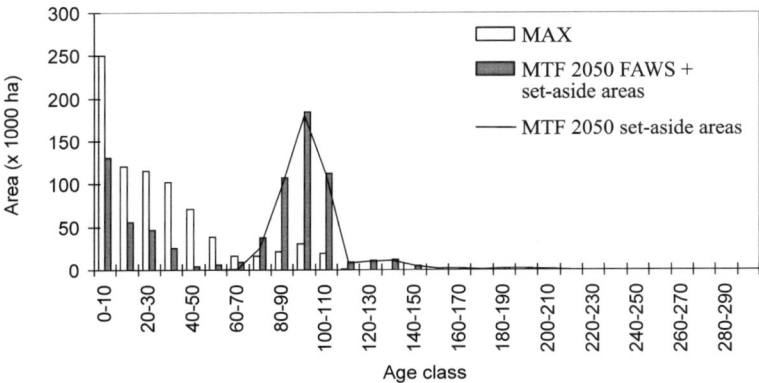

Fig. 5.49. The age class distribution of the Forest Available for Wood Supply (FAWS) in the initial situation and for 2050 under the BAU scenario (top). The age class distribution of FAWS for 2050 under the MAX scenario, the MTF scenario including set-aside areas and the MTF scenario with only set-aside areas (bottom).

Under the MTF scenario, the fellings first decrease, and then increase again to stabilise at 0.9 million m³/yr, leading to an average growing stock of 104 m³/ha by 2050. Under this scenario, mortality rates may then amount to 13% of the gross annual increment. The gross annual increment decreases in all scenarios from 2.4 m³/ha/yr initially, to 1.9 m³/ha/yr in 2010 due to the ageing of the coppice system.

The age class distribution for 2050 clearly shows the ageing process of the coppice system under the BAU scenario because the utilisation ratio is around 50–60% only. The ageing process occurs far less under the MAX scenario. In the MTF scenario 460 000 ha (60%) are projected to be newly established set-aside areas by 2050, leading to problems in finding the required fellings after 2040. This rather strong scenario of establishing reserves has led to an average build-up of growing stock in the forests of Macedonia, but has probably led to a decrease of growing stock in the managed part of the forest.

REFERENCES

Nilsson, S., Sallnäs, O. and Duinker, P. 1992. Future forest resources of Western and Eastern Europe. International Institute for Applied Systems Analysis. The Parthenon Publishing Group. England. 496 p.

UN-ECE/FAO 2000. Forest Resources of Europe, CIS, North America, Australia, Japan and New Zealand. Geneva, Timber and Forest Study Papers, No 17. 445 p.

5.18. THE NETHERLANDS

In co-operation with Henny Schoonderwoerd, Rino Jans and Douwe de Goede

Introduction

The small forest area of the Netherlands of 339 000 ha serves mainly as recreational space to the urbanised society (UN-ECE/FAO 2000). Most of the forests were established on former heathland or even further degraded sites. So, despite good growing conditions from the mild Atlantic climate, the growth of the current forest is limited. Forestry is carried out at a very small scale in the Netherlands, and many owners are not interested in revenues from their forests. The role of social functions, such as recreation, landscape and amenity are dominant. Nature conservation organisations are large owners in the Netherlands, putting emphasis on a conversion type of forestry leading to a closer-to-nature forest management (Schmidt et al. 1999). This is supported by the current forest policies.

Country specific scenario assumptions

The data underlying the projections covered 304 000 ha of forest, thus covering 97% of the forests available for wood supply according to UN-ECE/FAO (2000). The data are based on the 1986–1992 inventory cycle. The data were distinguished by 14 tree species.

Business-as-usual scenario

To initialise the required national production, several statistics were used. UN-ECE/FAO (2000) reports a production of 1.56 million m^3 of roundwood per year, of which 64% is from conifers. ETTS-V, which was initialised also for 1990, starts with annual fellings of 1.48 million m^3/yr. The COMTRADE removals data provide a slightly decreasing trend from 1.4 million m^3/yr in 1990 to 1.1 million m^3/yr in 1997. The projections were thus initialised with fellings of 1.5 million m^3/yr. Some 60% of total fellings from conifers was assumed to come from thinnings, for deciduous this was assumed to be 60% as well. It was assumed that the area of forest will stay at the 1990s level and that there is no change in species composition.

EFISCEN-ETTS scenario

ETTS-V foresaw an increasing trend in fellings until 2010, from 1.48 million m^3/yr in 1990 to 1.78 million m^3/yr in 2010 on 287 000 ha of exploitable forest. ETTS-V foresaw no forest area expansion.

Maximum sustainable production scenario

The fellings were quickly increased to arrive at a point where fellings equal increment; i.e. a level of 2.4 million m^3/yr by 2005.

Multi-functional scenario

The initial annual fellings of the BAU scenario was applied, and then it was assumed that the fellings will increase by 0.6% per year until 2010 to arrive at a level of 1.7 million m^3/yr. After that fellings stabilise. The rotation lengths were prolonged by 20 years. The share from thinnings were further increased to 70% from total fellings. The area initially over 110 years of oak and Scots pine, and over 80 years of alder were set aside as reserves. Initially this represents 21 000 ha. Through ageing this can increase during the simulation. Afforestation was assumed to take place at half the pace foreseen in the policy plans, i.e. 25 000 ha in total until 2020.

Results for The Netherlands

Most apparent from the projections for the Netherlands is the strong ageing of the forest under all scenarios. This is because the utilisation ratio in most scenarios is around 65% only, and because a large proportion of the fellings comes from thinnings. The ageing process is strongest in the MTF scenario because in the first scenario 56 000 ha are projected to be newly established set-aside areas by 2050. Under this MTF scenario, 70% of the fellings comes from thinnings, and a total production of dead wood of 0.22 million m^3/yr is achieved. The MTF scenario shows that this nature orientation in forest management can be combined with an increase in total annual fellings to around 1.63 million m^3/yr.

The projections for the 0.304 million ha of forests show that a maximum sustainable production of 1.77 million m^3/yr may be achieved, which is a moderate increase from the present level of around

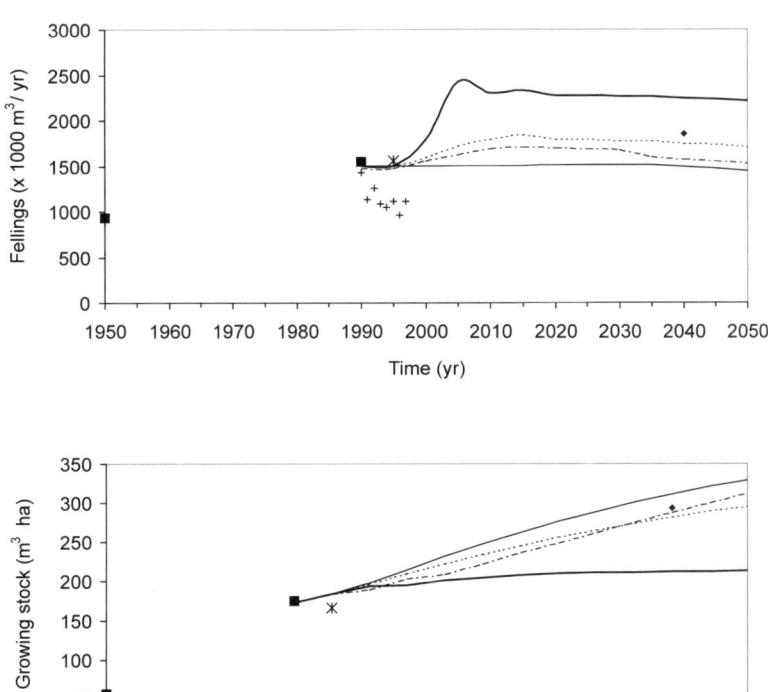

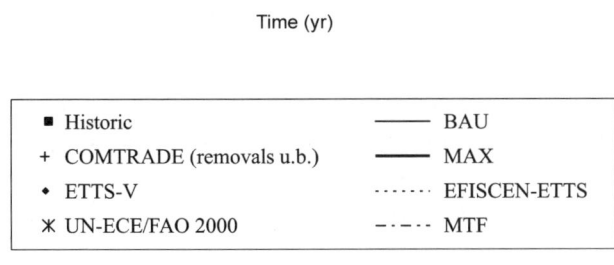

Fig. 5.50. Projected total national fellings (top) and growing stock per ha (bottom) in Dutch forests under the four scenarios until 2050.

1.5 million m^3/yr. By 2050 the growing stock under the MTF scenario may amount to 312 m^3/ha, whereas under the BAU scenario it may amount to 323 m^3/ha. Under the BAU scenario, mortality rates may then amount to 9.5% of the gross annual increment.

The gross annual increment shows approximately the same trend for all scenarios; an increase to around 8.7 m^3/ha/yr by 2015, and

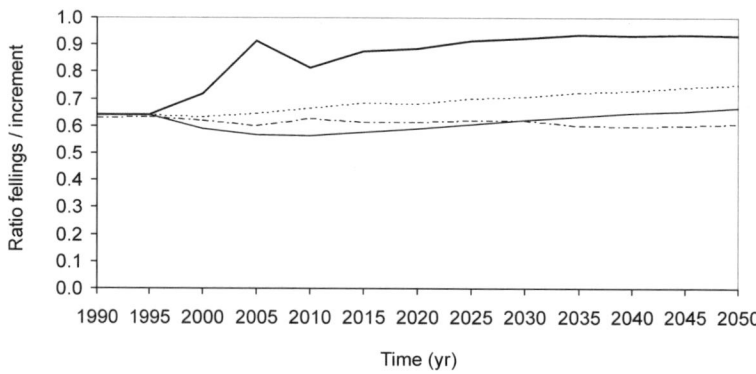

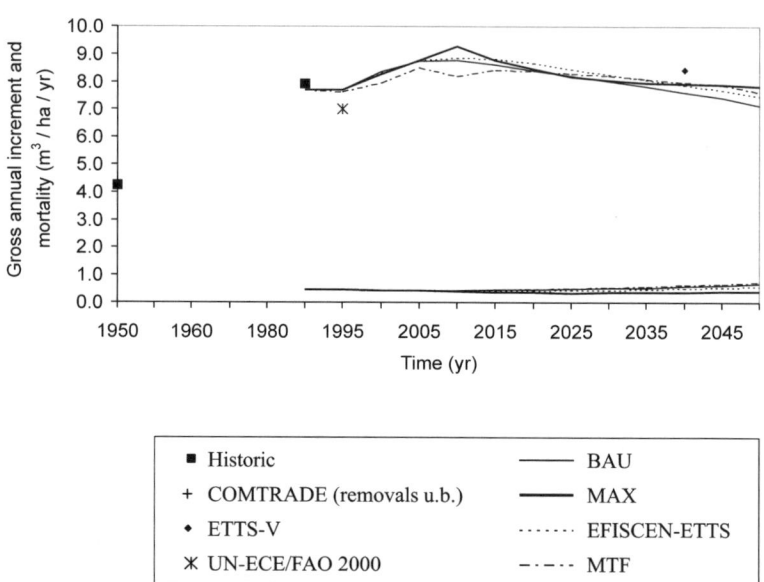

Fig. 5.51. Ratio of projected fellings/increment in Dutch forests under the four scenarios until 2050 (top) and projected net annual increment and mortality (bottom).

then a decline to, for example, 7.65 m³/ha/yr in 2050 in the MTF scenario. The decline is the strongest in the scenario where the build-up of growing stock is the largest: i.e. the BAU scenario with an increment of 7.15 m³/ha/yr in 2050.

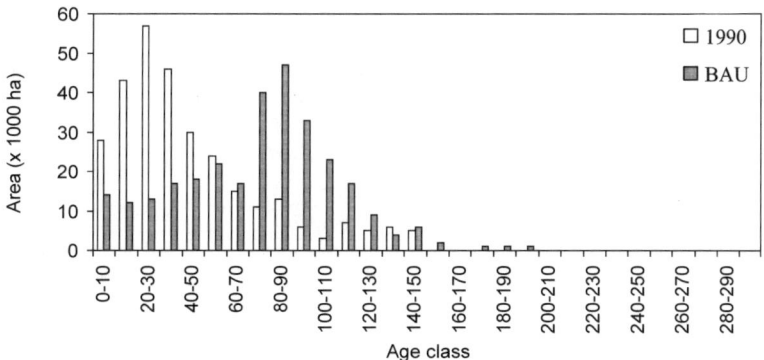

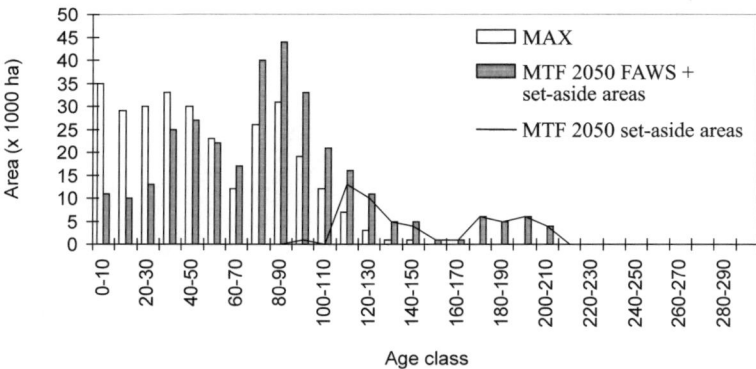

Fig. 5.52. The age class distribution of the Forest Available for Wood Supply (FAWS) in the initial situation and for 2050 under the BAU scenario (top). The age class distribution of FAWS for 2050 under the MAX scenario, the MTF scenario including set-aside areas and the MTF scenario with only set-aside areas (bottom).

REFERENCES

Schmidt, P., Kuiler, E., Wiersum, F. and Filius, B. 1999. The Netherlands. In: Pelkonen, P., Pitkänen, A., Schmidt, P., Oesten, G., Piussi, P. and Rojas, E. (eds.), Forestry in Changing Societies in Europe. Part II. SILVA Network. University Press, Joensuu, Finland. Pp. 229–254.

UN-ECE/FAO 2000. Forest Resources of Europe, CIS, North America, Australia, Japan and New Zealand. Geneva, Timber and Forest Study Papers, No 17. 445 p.

Further information

Edelenbosch, N.H. 1996. Wood harvest forecast model HOPSY. In: Päivinen, R., Roihuvuo, L. and Siitonen, M. (eds.), Large scale forestry scenario models: experiences and requirements. EFI Proceedings 5. European Forest Institute. Pp. 49–62.

5.19. NORWAY

In co-operation with Stein M. Tomter

Introduction

One of the major concerns in Norwegian forestry today is the fact that an increasing number of forest owners get most of their income from other types of activities than forestry. The reason for this is partly a reduced profitability from timber harvesting. An increasing percentage of the forest area has become unavailable for wood supply, for economic reasons. Forest owners are being encouraged to increase and develop their businesses based on non-wood goods and services, like rural tourism, hunting, fishing, and production of Christmas trees and decorative foliage. It is the Government's goal that the financial status of the forestry sector should be improved, while more attention has been focussed on the forest owners' responsibilities, guidance and information. The contribution of the forestry and forest sector industries to GDP is 2% (Svensrud 1999). Two important aims are to increase the degree of refinement of forestry products and to be able to significantly utilise all tree species.

Country specific scenario assumptions

The data underlying the projections covered 7.15 million ha of forest, while UN-ECE/FAO (2000) reported a forest area of 8.7 million ha and an area available for wood supply of 6.6 million ha thus covering 107% of the forests available for wood. The data are based on the 1986–1993 inventory cycle. The data were distinguished by 17 regions, eight site classes, and three species. Since broadleaved are usually mixed in other stands, the basic data underestimated the total growing stock for this species group. A correction was carried out for this.

Business-as-usual scenario

UN-ECE/FAO (2000) reported a production of 11.6 million m^3 of roundwood per year, of which 89% is from conifers. ETTS-V (Pajuoja 1995) starts with higher annual fellings of 13.5 million m^3/yr for 1990. The COMTRADE removals underbark data provide a sharp decreasing trend from 11.8 million m^3/yr in 1990 to 8.6 million m^3/yr in 1997. The projections were initialised with fellings according to ETTS-V: 12.2 million m³/yr from coniferous forests and 1.3 million m³/yr from deciduous forests. Some 10% of total fellings was assumed to come from thinnings. It was assumed that the area of forest will stay at the 1990s level and that there is no change in species composition.

EFISCEN-ETTS scenario

ETTS-V foresaw an gradual increasing trend in fellings to 15.1 million m³/yr up until 2010 without any changes in forest area.

Maximum sustainable production scenario

The fellings in deciduous forests could quickly be increased from 1.3 million m³/yr in 1990 to 5.0 million m³/yr in 2010 to arrive at a point where fellings equal increment. However, fellings in coniferous forests in 1990 were found to be at the maximum level at 12 million m³/yr.

Multi-functional scenario

For deciduous forests the trend in ETTS-V was followed, but for coniferous forests, the opposite trend was applied: decreasing from 12.2 to 11.6 million m³/yr in 2015. After that fellings stabilise. All rotation lengths were prolonged by 20 years. The share from thinnings were increased to 35% from total fellings. All forests initially over 180 years of age were set aside as reserves. Initially this represents 0 ha. Through ageing this can increase during the simulation. A very small area of forest expansion was assumed: 1% or 82 000 ha by 2015.

Results for Norway

The gross annual increment shows approximately the same trend for all scenarios; a decline to around 2.3 m³/ha/yr by 2050. The decline is the least in the MTF scenario, because in this latter scenario a larger proportion of fellings originates from thinnings. ETTS-V still foresaw increment to increase to 4.9 m³/ha/yr by 2050. The decline found here causes the fellings-increment ratio to rise steadily, despite rather stable total fellings levels.

The projections for 7.1 million ha of forests show that a maximum sustainable production of 15 million m³/yr may be achieved, which is only a small increase from the present level. However, this rather low level is mainly caused by the declining gross annual increment. This is

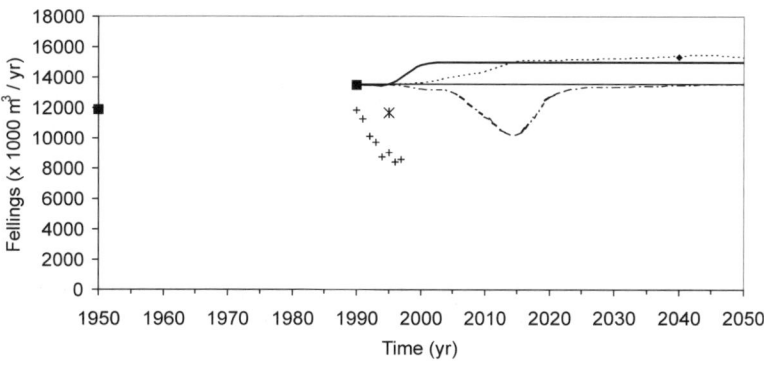

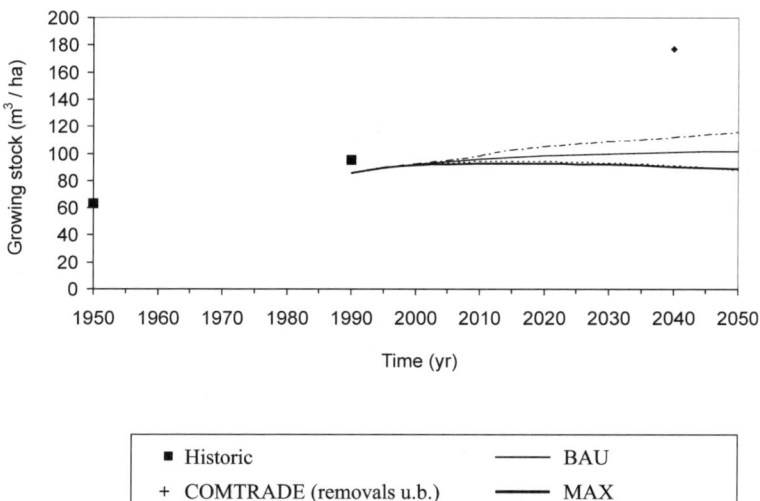

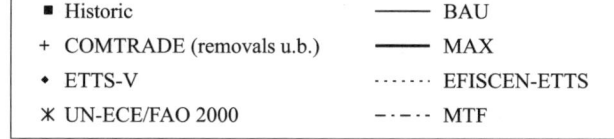

Fig. 5.53. Projected total national fellings (top) and growing stock per ha (bottom) in Norwegian forests under the four scenarios until 2050.

again caused by harvesting a large proportion of total fellings through final fellings; thus affecting gross annual increment in a negative way.

By 2050 the growing stock under the MAX scenario has remained at 89 m^3/ha, whereas under the MTF it may amount to 117 m^3/ha. Under this latter scenario, mortality rates may then amount to 16% of the gross annual increment.

142　　　　　　　　　　COUNTRY LEVEL RESULTS

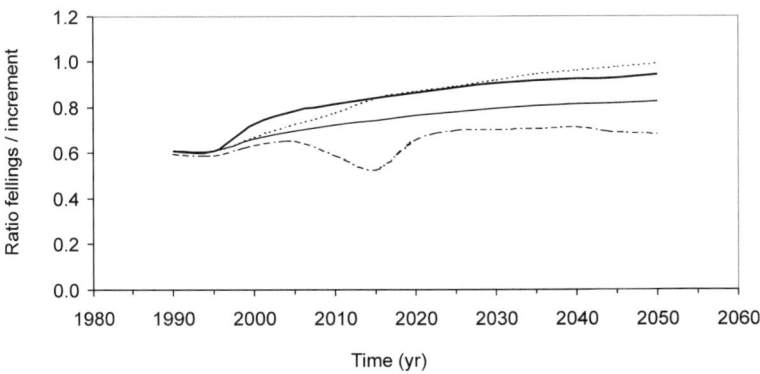

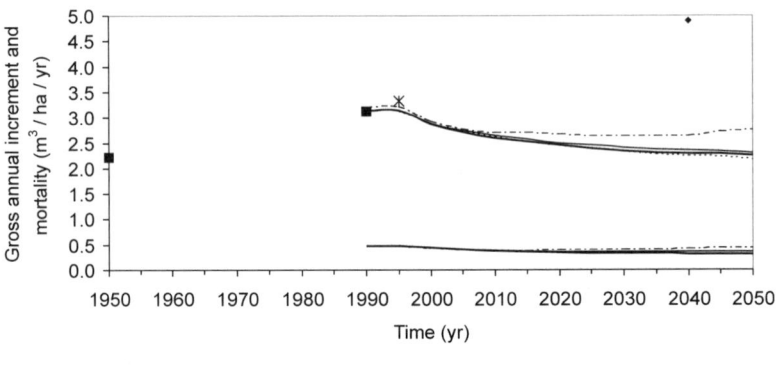

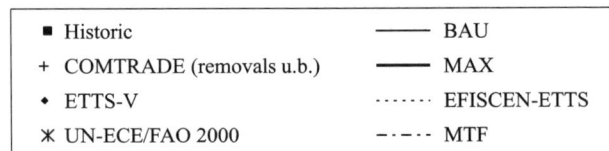

Fig. 5.54. Ratio of projected fellings/increment in Norwegian forests under the four scenarios until 2050 (top) and projected net annual increment and mortality (bottom).

The projections for the Norwegian forests age class distributions show almost no ageing of the forest under all scenarios. This is because the utilisation ratio in most scenarios is rather high (mostly above 75%) and because a large proportion of the fellings comes from final fellings. A total production of dead wood of 3.2 million m³/yr is achieved in the

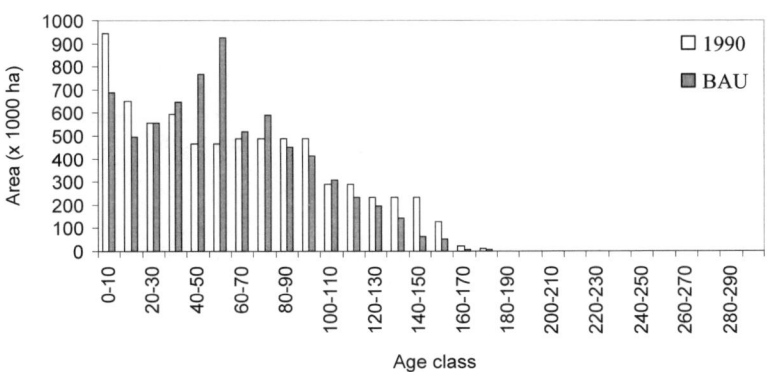

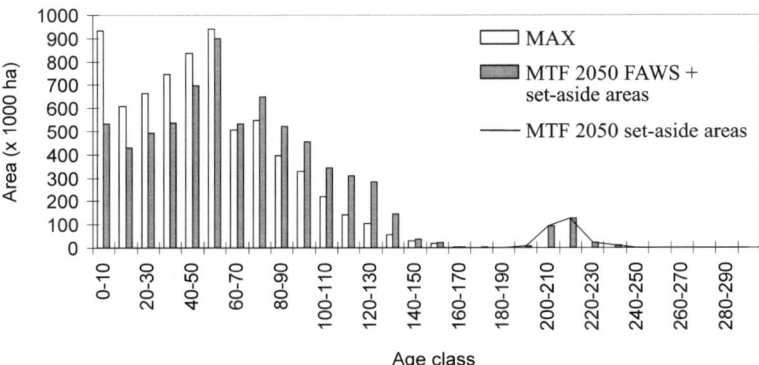

Fig. 5.55. The age class distribution of the Forest Available for Wood Supply (FAWS) in the initial situation and for 2050 under the BAU scenario (top). The age class distribution of FAWS for 2050 under the MAX scenario, the MTF scenario including set-aside areas and the MTF scenario with only set-aside areas (bottom).

MTF scenario. The MTF scenario shows the rather narrow possibilities in Norwegian forests; fellings could not be found between 2005 and 2025 due to a combined effect of declining increment, set-aside, and prolonging rotations.

REFERENCES

Pajuoja, H. 1995. The outlook for the European Forest Resources and roundwood supply. Geneva, Timber and Forest Discussion Papers ECE/TIM/DP/4. UN-ECE/FAO. Geneva. 59 p.

Svensrud, A. 1999. Norway. In: Pelkonen, P., Pitkänen, A., Schmidt, P., Oesten, G., Piussi, P. and Rojas, E. (eds.), Forestry in Changing Societies in Europe. Part II. SILVA Network. University Press, Joensuu, Finland. Pp. 255–273.

UN-ECE/FAO 2000. Forest Resources of Europe, CIS, North America, Australia, Japan and New Zealand. Geneva, Timber and Forest Study Papers, No 17. 445 p.

Further information

Hoen, H. 1996. Forestry scenario modelling for economic analysis – experiences using the GAYA-JLP model. In: Päivinen, R., Roihuvuo, L. and Siitonen, M. (eds.), Large scale forestry scenario models: experiences and requirements. EFI Proceedings 5. European Forest Institute. Pp. 79–88.

5.20. POLAND

In co-operation with Dr. Jan Glaz and Dr. Jerzy Wawrzoniak

Introduction

Forests cover 30% of Poland, occupying an area of 9 million ha. As much as 83% of the forests are in public ownership. Coniferous forest make up 77% of the total forest area. Polish forests are characterised by a contrast between the lowland and upland areas. Sanitary fellings make up a large proportion of the total fellings because natural mortality accounts for 28% of the gross increment (UN-ECE/FAO 2000, Kaczmarek and Dudek 1999). In 1995, the Council of Ministry adopted a programme of afforestation of 700 000 ha of poor quality arable lands until 2020 and for the further future a total of 1.5 million ha is aimed at. In the state forests, various types of reserves are maintained, in total covering 135 000 ha. Protective functions prevail on 48% of the forest area.

Country specific scenario assumptions

The data underlying the projections covered 6.8 million ha of forest, while UN-ECE/FAO (2000) reported a forest area of 8.9 million ha. The results presented here thus covered the state owned forest available for wood supply only (=82% of all forest available for wood supply). The data are based on the 1993 inventory and were distinguished by 17 regions and ten tree species.

Business-as-usual scenario

UN-ECE/FAO (2000) reported a production of 31.6 million m^3 of roundwood per year, of which 76% is from conifers. The COMTRADE removals underbark data show an increasing trend from 17.8 million m^3/yr in 1990 to 21.6 million m^3/yr in 1997. UN-ECE/FAO (1992) reported a coniferous production of 20.8 million m^3/yr and a production of deciduous wood of 6.5 million m^3/yr. The projections were initialised according to the latter figures taking into account that we cover 82% of all forest available for wood supply. Some 50% of total fellings was assumed to come from thinnings. It was assumed that the area of forest will stay at the 1990s level and that there is no change in species composition.

EFISCEN-ETTS scenario

ETTS-V foresaw a slow growth in fellings until 2040 of 0.3% per year; totalling 27.2 million m^3/yr in 2040. There was a forest area expansion of 670 000 ha until 2040.

Maximum sustainable production scenario

The fellings in coniferous forests could quickly be increased to 36 million m^3/yr in 2010 (on the 82% of the forests available for wood supply) to arrive at a point where fellings equal increment. Fellings in deciduous forests were increased to 6.4 million m^3/yr in 2010.

Multi-functional scenario

It was assumed that the fellings will first decrease by 5% (representing economic restrictions) and then remain at a rather low level until 2015, and only then start to increase (representing economic growth) by 2.5% per 5 year period until 2050. All rotation lengths were prolonged by 20 years. The share from thinnings was decreased to 40% from total fellings. All forests of beech, oak, hornbeam, and alder that were over 120 years of age at the start were set aside as reserves. Initially this represents 152 000 ha. Through ageing this can increase during the simulation. Forest area expansion was assumed to be successful for about two-thirds of the area mentioned in the EFISCEN-ETTS scenario.

Results for Poland

The results for Poland show large biological potential for increasing the fellings. On the simulated 82% of the forests available for wood supply, the maximum sustainable production scenario shows that a production of 42 million m^3/yr may be achieved; almost double the present level. The gross annual increment shows approximately the same trend for all scenarios; rather stable until 2020, and then declining slightly to 5.8 m^3/ha/yr in 2050 due to ageing and increase in growing stock. Based on decline phenomena of that time, ETTS-V foresaw an increment decline to 3.1 m^3/ha/yr by 2050.

By 2050 the growing stock under the MAX scenario has remained at 210 m^3/ha (compared with 194 m^3/ha in 1990), whereas under

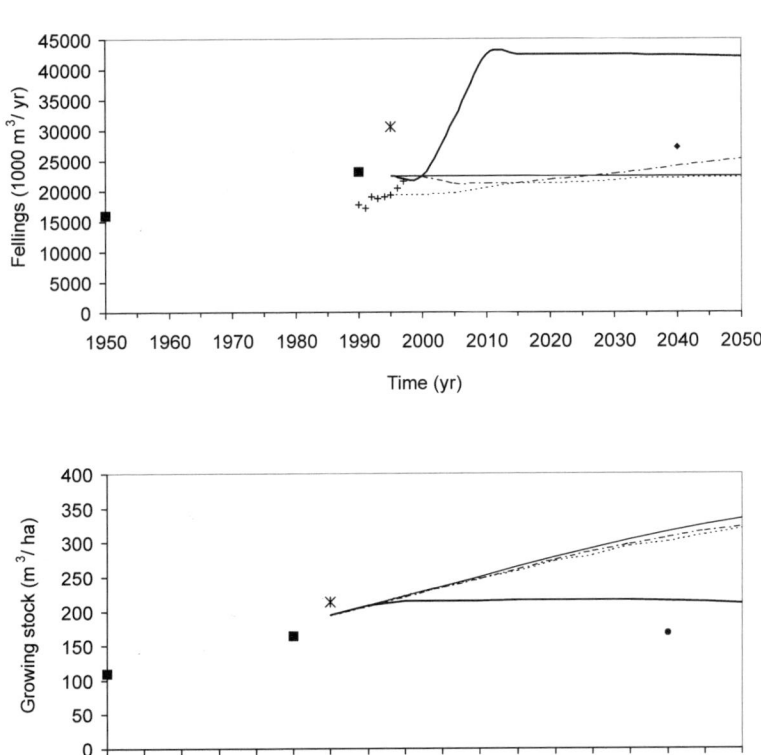

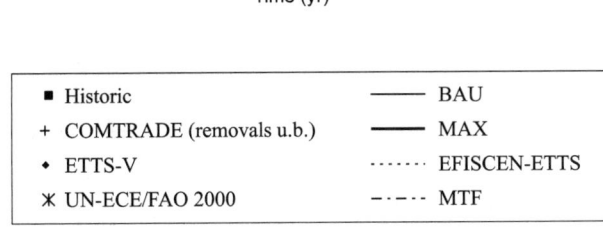

Fig. 5.56. Projected total national fellings (top) and growing stock per ha (bottom) in Polish forests under the four scenarios until 2050.

the MTF it may amount to 322 m^3/ha. Under this latter scenario, mortality rates may then amount to 14% of the gross annual increment.

The age class distributions show a clear ageing of the forest under all scenarios, except for the MAX scenario. This is because the utilisation ratio in most scenarios is low: mostly slightly below 50%. This results

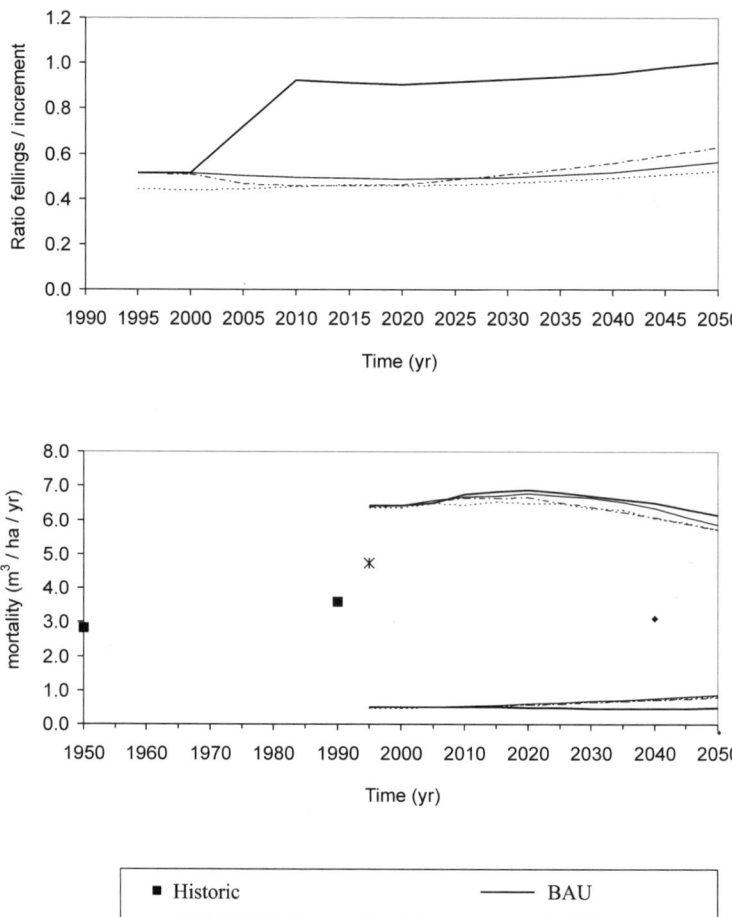

Fig. 5.57. Ratio of projected fellings/increment in Polish forests under the four scenarios until 2050 (top) and projected net annual increment and mortality (bottom).

in small regeneration areas, and thus small areas of forest in the young and middle age classes in 2050.

A total production of dead wood of 5.7 million m³/yr is achieved in the MTF scenario. The MTF scenario shows ample possibilities in Polish forests for nature-oriented management; a fellings increase

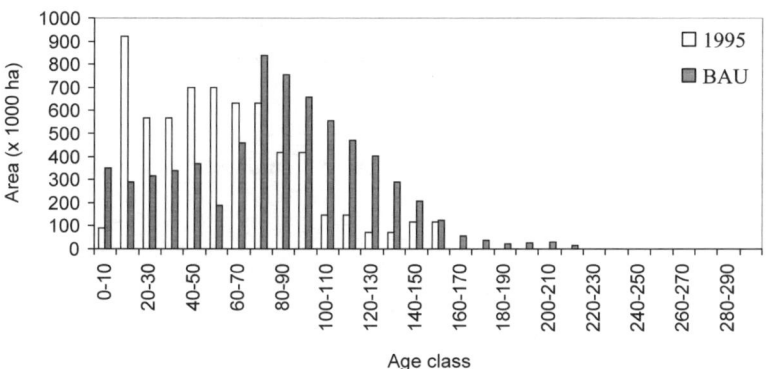

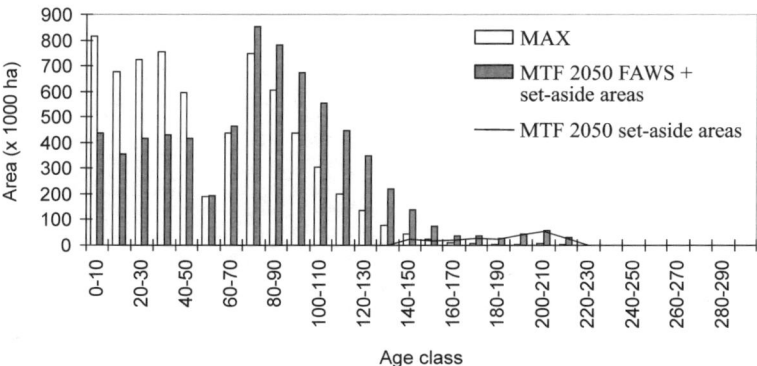

Fig. 5.58. The age class distribution of the Forest Available for Wood Supply (FAWS) in the initial situation and for 2050 under the BAU scenario (top). The age class distribution of FAWS for 2050 under the MAX scenario, the MTF scenario including set-aside areas and the MTF scenario with only set-aside areas (bottom).

of 2.5% per 5 years was possible while prolonging the rotation, and having set aside 323 000 ha for reserves in 2050.

REFERENCES

Kaczmarek, K. and Dudek, A. 1999. Poland In: Pelkonen, P., Pitkänen, A., Schmidt, P., Oesten, G., Piussi, P. and Rojas, E. (eds.), Forestry in Changing Societies in Europe. Part II. SILVA Network. University Press, Joensuu, Finland. Pp. 275–293.

UN-ECE/FAO 1992. The Forest Resources of the Temperate Zones, the UN-ECE/FAO 1990 Forest Resource Assessment, General Forest Information. Geneva, Switzerland. 347 p.

UN-ECE/FAO 2000. Forest Resources of Europe, CIS, North America, Australia, Japan and New Zealand. Geneva, Timber and Forest Study Papers, No 17. 445 p.

Further information

Central Statistical Office. 1997. Forestry 1997 – information and statistical papers. Warszawa. 265 p.

5.21. PORTUGAL

In co-operation with Dr. A. Leite

Introduction

Portugal has both Atlantic as well as Mediterranean climatic conditions. Of the 3.3 million ha of forest, 85% is in private ownership. The main species are *Pinus pinaster*, cork oak and eucalypts. Large areas of forests are managed as plantations. Thus, despite a moderate national forest area, the annual fellings are considerable. The forest cluster contributes 2.6% to the GDP, making it an important sector although fellings are usually difficult and costly. Threats to this contribution to the GDP are forest fires and a sharply increasing demand for wood products, while production and afforestation are not keeping up with that (Carvalho Mendes 1999).

Country specific scenario assumptions

The data underlying the projections covered 1.6 million ha of forest, while UN-ECE/FAO (2000) reported a forest area of 3.3 million ha, of which 1.9 million ha is available for wood supply. The results presented here thus covered all of the coniferous forests and 78% of the broadleaved forests available for wood supply. The felling levels were reduced accordingly in all scenarios. The data are based on the 1992 inventory and were distinguished by four regions and two tree species (*Pinus pinaster* and *Eucalyptus* spp. each by four dominance classes).

Business-as-usual scenario

UN-ECE/FAO (2000) reported a production of 11.5 million m^3 of roundwood per year, approximately equally distributed between coniferous and broadleaved species. The COMTRADE removals underbark data show a decreasing trend from 11.3 million m^3/yr in 1990 to 9 million m^3/yr in 1997. UN-ECE/FAO (1992) reported a coniferous production of 6.6 million m^3/yr and a production of broadleaved wood of 4.3 million m^3/yr. The projections were initialised according to the latter figures taking into account that we cover 78% of broadleaved forest available for wood supply. Some 25% of total coniferous fellings was assumed to come from thinnings, in broadleaved forests no thinnings were assumed (as eucalypts are managed in a coppice system). It was assumed that the area of forest will stay at the 1990s level and that there is no change in species composition.

EFISCEN-ETTS scenario

ETTS-V foresaw slightly increasing fellings until 2040 of 0.3% per year in broadleaved forests and 0.1% in coniferous forests. This comes to a total of 11.5 million m^3/yr in 2040 for the proportion of the forest covered here. ETTS-V foresaw forest area expansion of 380 000 ha until 2040. It was assumed that there is no change in species composition.

Maximum sustainable production scenario

Since the fellings level of the EFISCEN-ETTS and the BAU scenarios were found to be unsustainable, the fellings in coniferous forests were decreased to 4.4 million m^3/yr (on this proportion of the forests available for wood supply) to arrive at a point where fellings equal increment. Fellings in broadleaved forests were increased to 6.4 million m^3/yr in 2010. Production of broadleaved wood was kept at 4.1 million m^3/yr.

Multi-functional scenario

Since the MAX scenario was found to be the only sustainable scenario, these fellings levels were used here as well. All rotation lengths were prolonged by 20 years. The share from thinnings remained at 25% from total fellings in coniferous forests. All forests of mixed stands of *Pinus pinaster* initially over 60 years of age were set aside as reserves. Initially this represents 69 000 ha. Through ageing this can increase during the simulation. Forest area expansion was assumed to be successful for about half of the area mentioned in the EFISCEN-ETTS scenario.

Results for Portugal

The Portuguese results have to be regarded with care, because a rather small proportion of the total forest area is covered (48%) in these simulations. The provisional results as presented here show no potential for increasing the fellings. The results indicate that the maximum sustainable production level may be below the current fellings levels.

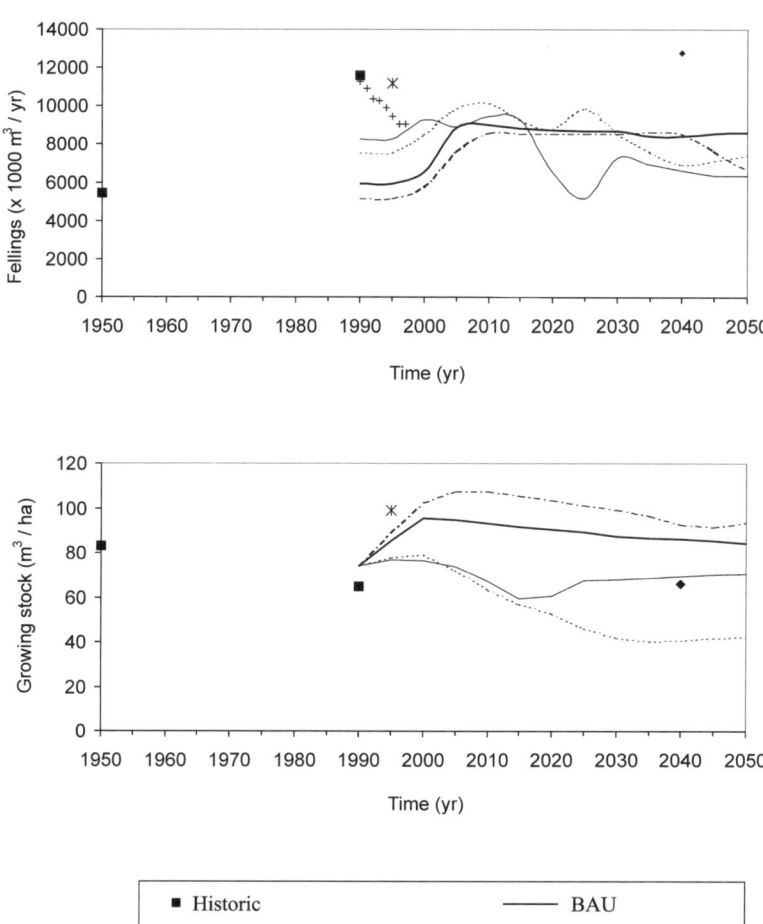

Fig. 5.59. Projected total national fellings (top) and growing stock per ha (bottom) in Portuguese forests under the four scenarios until 2050.

It shows that the production may have to be reduced to 5.9 million m³/yr at least until 2005, in order to build up growing stock. After that fellings can increase. The BAU scenario shows that the current fellings levels cannot be maintained, and the fellings thus drop sharply around 2020–2025. Fellings in the EFISCEN-ETTS scenario also resulted in a decline of growing stock: to 43 m³/ha in 2050. However, ETTS-V

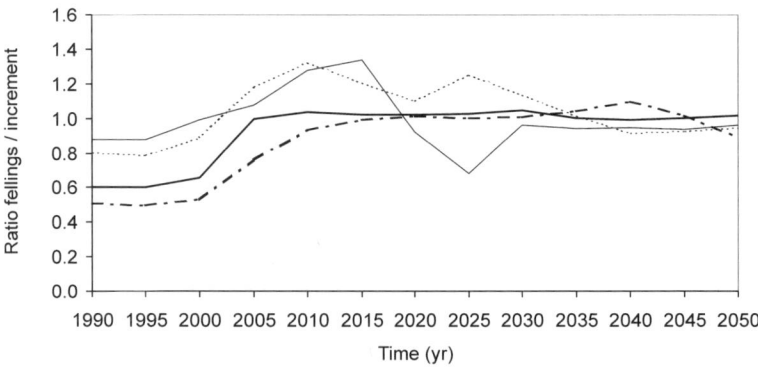

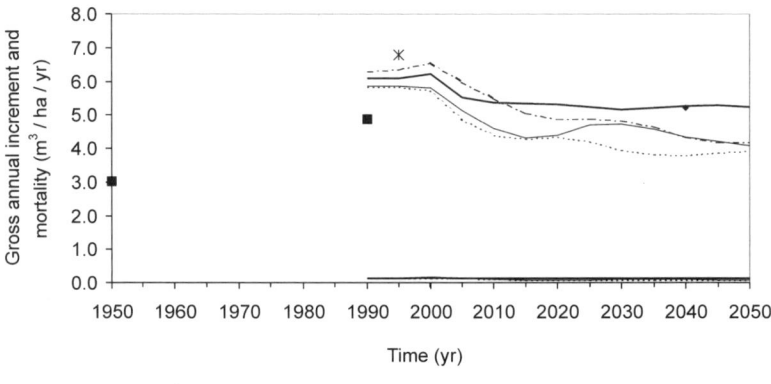

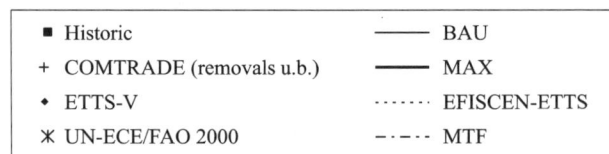

Fig. 5.60. Ratio of projected fellings/increment in Portuguese forests under the four scenarios until 2050 (top) and projected net annual increment and mortality (bottom).

had assumed a higher increment to remain under their felling level: 5.2 m³/ha/yr in 2040 (against 4.3 m³/ha/yr in our simulations). They also had a larger area included in their assessment: 2.3 million ha.

The gross annual increment shows a declining trend for all scenarios, although differing in extent. The least decline is reached

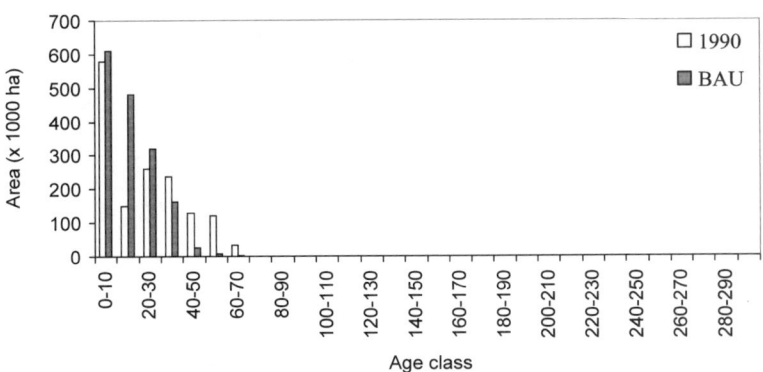

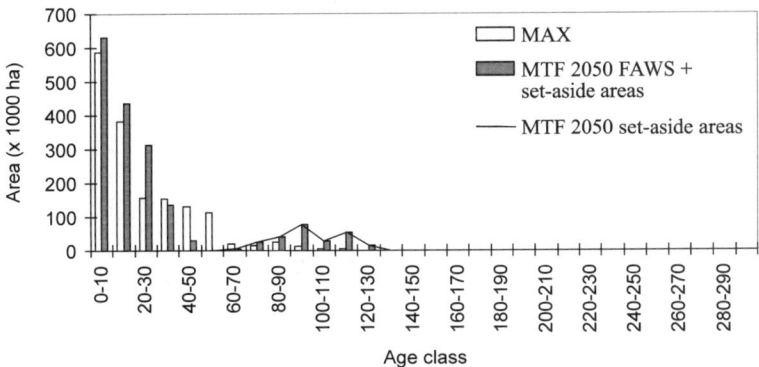

Fig. 5.61. The age class distribution of the Forest Available for Wood Supply (FAWS) in the initial situation and for 2050 under the BAU scenario (top). The age class distribution of FAWS for 2050 under the MAX scenario, the MTF scenario including set-aside areas and the MTF scenario with only set-aside areas (bottom).

in the MAX scenario where the growing stock is under less pressure. In this scenario, the increment declines from 6.1 m^3/ha/yr in 1990 to 5.3 m^3/ha/yr in 2050.

In the MTF scenario, most build-up of growing stock is reached: from 74 m^3/ha in 1990, peaking at 107 m^3/ha in 2010, and then declining to 93 m^3/ha in 2050. Under this scenario, mortality rates may then amount to 3% of the gross annual increment.

The age class distributions show almost no ageing of the forest under all scenarios, except for the MTF scenario. This is because the utilisation ratio in most scenarios is very high: mostly around 100%.

This results in large regeneration areas, and consequently large areas of forest in the young age classes in 2050.

On the 1.6 million ha of forests, a total production of dead wood of 0.22 million m^3/yr is achieved in the MTF scenario. The MTF scenario shows reasonable possibilities for nature-oriented management (under moderate fellings levels); having set aside 248 000 ha for reserves in 2050.

REFERENCES

Carvalho Mendes, M.S. 1999. Portugal. In: Pelkonen, P., Pitkänen, A., Schmidt, P., Oesten, G., Piussi, P. and Rojas, E. (eds.), Forestry in Changing Societies in Europe. Part II. SILVA Network. University Press, Joensuu, Finland. Pp. 295–322.

UN-ECE/FAO 1992. The Forest Resources of the Temperate Zones, the UN-ECE/FAO 1990 Forest Resource Assessment, General Forest Information. Geneva, Switzerland. 347 p.

UN-ECE/FAO 2000. Forest Resources of Europe, CIS, North America, Australia, Japan and New Zealand. Geneva, Timber and Forest Study Papers, No 17. 445 p.

Further information

Direcção-Geral das Florestas 2001 Inventario Florestal Nacional Portugal Continental, 3rd revision 1995–1998. Direccao-Geral das Florestas, Lisbon. 233 p.

5.22. ROMANIA

In co-operation with Mr. Sorin Popescu

Introduction

Romania's continental and rather dry climate of the plains is interrupted by moist temperate and alpine conditions in the Carpathian Mountains. Of the 6.3 million ha of forest (mostly in the mountains), less than half is managed as Group II forests – 'forests with productive and protective functions' (Milescu 1999). The rest has mainly a protective function. Still, UN-ECE/FAO (2000) reports 5.6 million ha as 'available for wood supply'. Beech and Norway spruce are the main tree species. Less than 5% of the forest is in private ownership, and the National Forest Administration is the main body allowed to sell wood. Although this administration still has a monopoly, it released prices of wood in the rough recently; one further step towards a market economy. Problems facing the forestry sector include: very slow privatisation, low investment levels, and low international trade activities.

Country specific scenario assumptions

The data underlying the projections covered 6.21 million ha of forest, while UN-ECE/FAO (2000) reported a forest area of 6.3 million ha of which 5.6 million ha is available for wood supply. The results presented here thus covered almost all of the Romanian forest area relevant in this context. The data are based on the early 1980s inventory (Nilsson et al. 1992) and were distinguished by five site classes and ten tree species. Some 0.56 million ha was under coppice management.

Business-as-usual scenario

UN-ECE/FAO (2000) reported a production of 13.6 million m^3 of roundwood per year, of which 63% is from deciduous species. The COMTRADE removals underbark data show a rather stable level around 12.6 million m^3/yr. UN-ECE/FAO (1992) reported a coniferous production of 10.5 million m^3/yr and a production of deciduous wood of 5.4 million m^3/yr. The projections were initialised according to the latter figures, with 2.1 million m^3 being harvested in coppice systems. Some 35% of total coniferous fellings was assumed to come from thinnings, in deciduous forest the figure was 30%, and in coppice it was 20%. It was assumed that the area of forest will stay at the 1990s level without changes in species distribution.

EFISCEN-ETTS scenario

ETTS-V foresaw stable fellings until 2040 and is thus the same scenario as the BAU. ETTS-V assumed that there is no forest area expansion and no species change.

Maximum sustainable production scenario

Substantial felling increases were possible in coniferous and deciduous high forests. The fellings were quickly increased to arrive at a point where fellings equal increment; i.e. a level of 43 million m^3/yr in high forests by 2005. In coppice systems, the fellings were slightly decreased to 1.8 million m^3/yr.

Multi-functional scenario

In the initial period, fellings were started with the same levels as given in the BAU scenario. Then it was assumed that due to the economic crisis, fellings drop to the levels given by UN-ECE/FAO (2000). As of 2005 it was assumed that fellings will increase by 2.5% per 5 years until 2020 in the high forests. After that, they stabilise. To represent the emphasis on nature-oriented management, all rotation lengths were prolonged by 20 years. The share from thinnings was increased to 40% in high forest, and 30% in coppice. All high forests of oak, beech and other broadleaves initially over 100 years of age were set aside as reserves. Also, all coppice systems of oak, beech and other hardwood initially over 50 years of age were set aside as reserves. Initially this represents an area of 684 000 ha. Through ageing this can increase during the simulation. Forest area expansion was assumed to be carried out to a small extent: 250 000 ha by 2015.

Results for Romania

All aspects of the results for Romanian forests show the present low level of fellings, and thus the large potential for increasing the fellings. The results indicate that the maximum sustainable production level can be maintained at 43 million m^3/yr (compared with 13 million

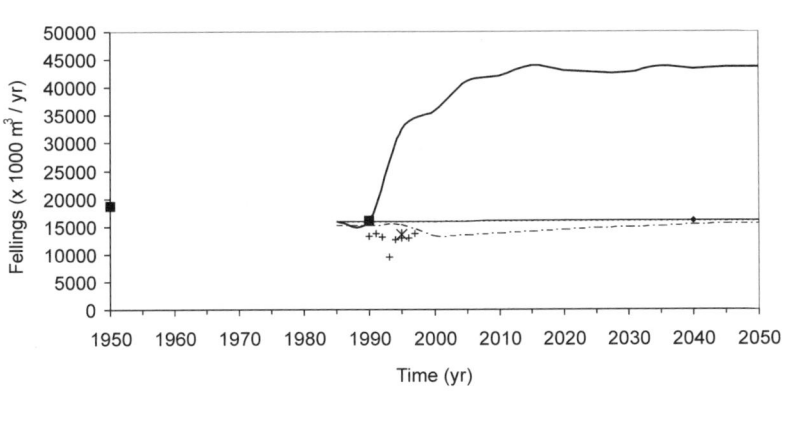

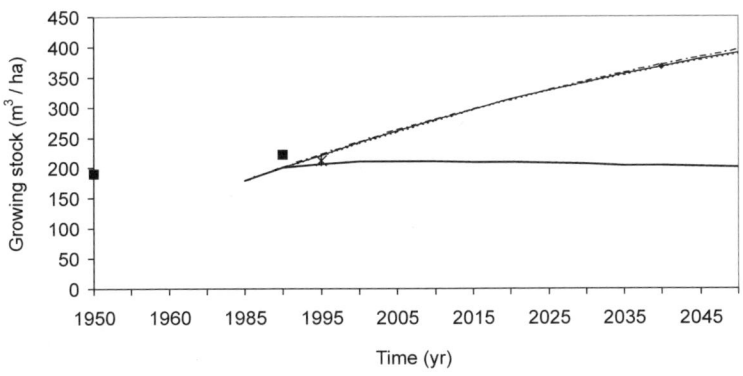

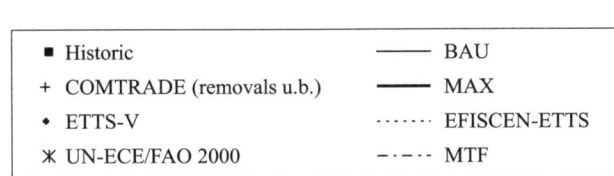

Fig. 5.62. Projected total national fellings (top) and growing stock per ha (bottom) in Romanian forests under the four scenarios until 2050.

m^3/yr at present). Under the BAU scenario the growing stock may develop to 389 m^3/ha in 2050.

The gross annual increment shows a slightly declining trend for all scenarios, except the MAX. In the other scenarios, increment declines from 7.3 m^3/ha/yr in 1990 to 5.6 m^3/ha/yr in 2050. This is due to ageing and the build-up of large growing stocks.

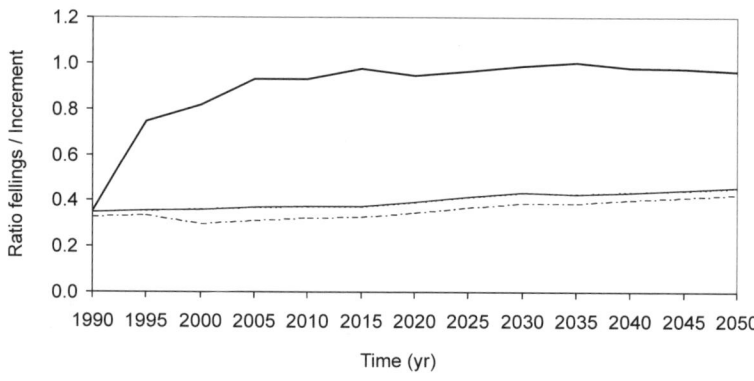

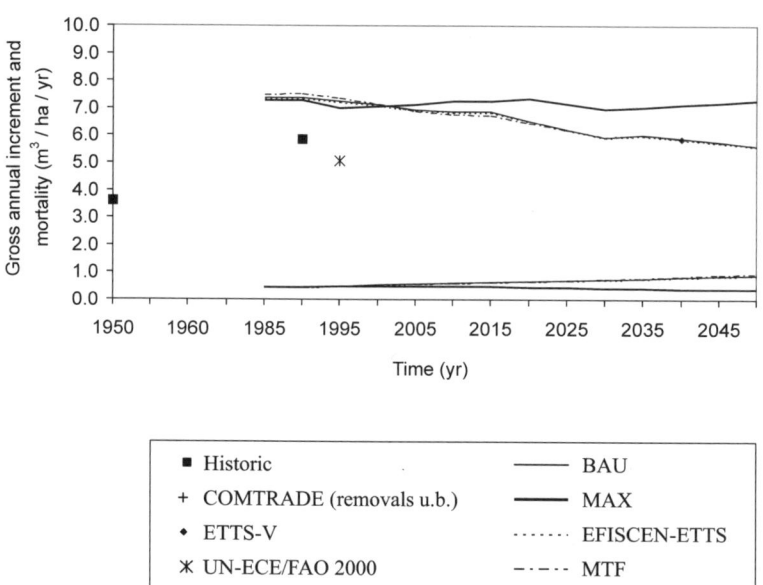

Fig. 5.63. Ratio of projected fellings/increment in Romanian forests under the four scenarios until 2050 (top) and projected net annual increment and mortality (bottom).

In the MTF scenario, most of the increase in growing stock is reached because an intermediate period with even lower fellings than in the BAU scenario was assumed: 395 m³/ha in 2050. Under this latter scenario, mortality rates may then amount to 17% of the gross annual increment.

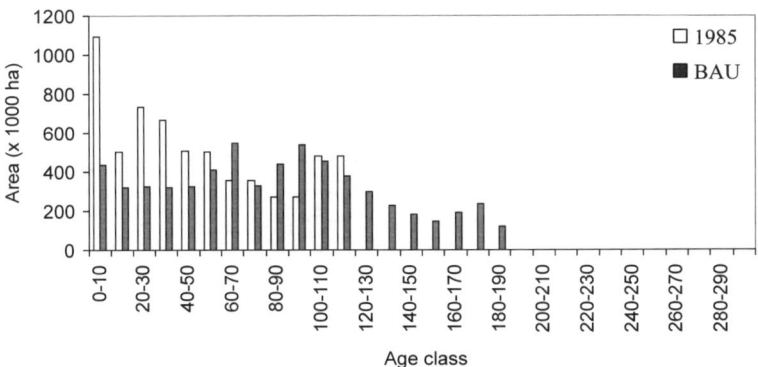
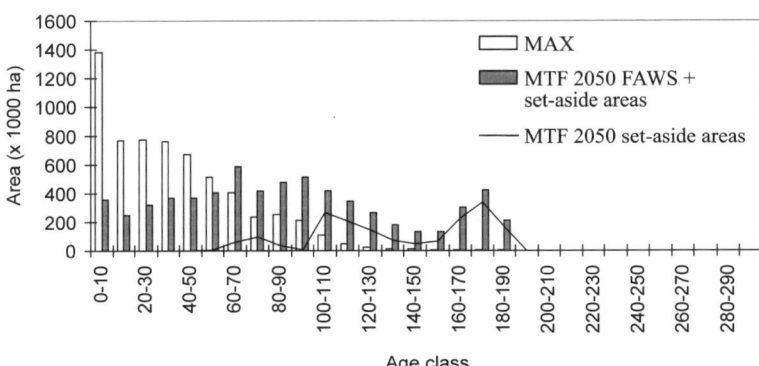

Fig. 5.64. The age class distribution of the Forest Available for Wood Supply (FAWS) in the initial situation and for 2050 under the BAU scenario (top). The age class distribution of FAWS for 2050 under the MAX scenario, the MTF scenario including set-aside areas and the MTF scenario with only set-aside areas (bottom).

The age class distributions show a strong ageing of the forest under all scenarios, except for the MAX scenario. This is because the utilisation ratio in most scenarios is very low. This results in small regeneration areas, and consequently small areas of forest in the young and middle age classes in 2050.

The MTF scenario shows ample possibilities for nature-oriented management under these low fellings levels. By 2050, 1.74 million ha of forests were set aside as strict reserves, causing no problems in finding the required fellings.

REFERENCES

Milescu, I. 1999. Romania. In: Pelkonen, P., Pitkänen, A., Schmidt, P., Oesten, G., Piussi, P. and Rojas, E. (eds.), Forestry in Changing Societies in Europe. Part II. SILVA Network. University Press, Joensuu, Finland. Pp. 323–334.

Nilsson, S., Sallnäs, O. and Duinker, P. 1992. Future forest resources of Western and Eastern Europe. International Institute for Applied Systems Analysis. The Parthenon Publishing Group. England. 496 p.

UN-ECE/FAO 1992. The Forest Resources of the Temperate Zones, the UN-ECE/FAO 1990 Forest Resource Assessment, General Forest Information. Geneva, Switzerland. 347 p.

UN-ECE/FAO 2000. Forest Resources of Europe, CIS, North America, Australia, Japan and New Zealand. Geneva, Timber and Forest Study Papers, No 17. 445 p.

5.23. SLOVAK REPUBLIC

In co-operation with Dr. Ivan Luptak

Introduction

Much of the Slovakian forests are situated in the western part of the Carpathians. In total forests cover 40% of the land area of Slovakia. Of the 2 million ha, 43% is state owned, and 21% community owned (UN-ECE/FAO 2000). Coniferous, broadleaved and mixed forests each comprise about one-third of the total forest area (Scheer and Longauer 1999). Forest health remains a concern in Slovakian forestry. About half of the total fellings in Slovakia (7.4 million m^3/yr) still consists of salvaged wood. Much of the harvesting is carried out in a small-scale shelterwood system. This will probably increase under the envisaged increase in nature-oriented forest management. The forest sector is important to the Slovakian economy with a contribution of 7.4% to the GDP.

Country specific scenario assumptions

The data underlying the projections covered 1.82 million ha of forest, while UN-ECE/FAO (2000) reported a forest area available for wood supply of 1.7 million ha. The data are based on the 1994 inventory, but for increment the data from Nilsson et al. (1992) were used. The data were distinguished by four degrees of protection, and by 21 tree species.

Business-as-usual scenario

UN-ECE/FAO (2000) reported a production of 7.4 million m^3 of roundwood per year, of which two-thirds is from conifers. The COMTRADE removals underbark data show an increasing trend from 5.2 million m^3/yr in 1993 to 5.9 million m^3/yr in 1997. The projections were initialised according to UN-ECE/FAO (2000). Some 30% of total fellings was assumed to come from thinnings. It was assumed that the area of forest will stay at the 1990s level and that there is no change in species composition.

EFISCEN-ETTS scenario

ETTS-V foresaw a decline in fellings until 2000, then a gradual increase until 2030, and then again a sharp fall of 17%; amounting to 6.1 million m^3/yr in 2040. ETTS-V foresaw a forest area expansion of 14 000 ha by 2010.

Maximum sustainable production scenario

Fellings in both coniferous and deciduous forests were increased to 10.2 million m^3/yr in 2005 to arrive at a point where fellings equal increment.

Multi-functional scenario

It was assumed that the fellings decline and rise as foreseen by ETTS-V may be realistic. This trend was thus followed, although we foresee a slower increase (8% per 10 years) in the period 2005–2030. To represent the desire for nature-oriented forest management, all rotation lengths were prolonged by 10 years and the share of thinnings from total fellings was increased to 50%. All forests initially of linden, oak, Turkey oak, and beech over 130 years of age were set aside as reserves. Initially this represents an area of 28 000 ha. Forest area expansion was assumed to amount to 80 000 ha by 2035, and was equally distributed between poplar and Norway spruce.

Results for Slovak Republic

The results for Slovakia are dominated by a wide range in assumptions on future fellings. The results show large biological potential for increasing the fellings; on the simulated forests available for wood supply the MAX scenario shows that a production of 10.2 million m^3/yr may be achieved. However, the uncertainty over economic recovery are the basis for the assumption for further decline in fellings, and recovery only later on in the period 2005–2030.

The low utilisation level under the MTF scenario leads to a further build-up of growing stock; by 2050 it may amount to 319 m^3/ha. Under this latter scenario mortality rates may then amount to 18% of the gross annual increment. ETTS-V, which foresaw a much more moderate gross annual increment, thus also resulted in a much lower growing stock by 2050 of 185 m^3/ha.

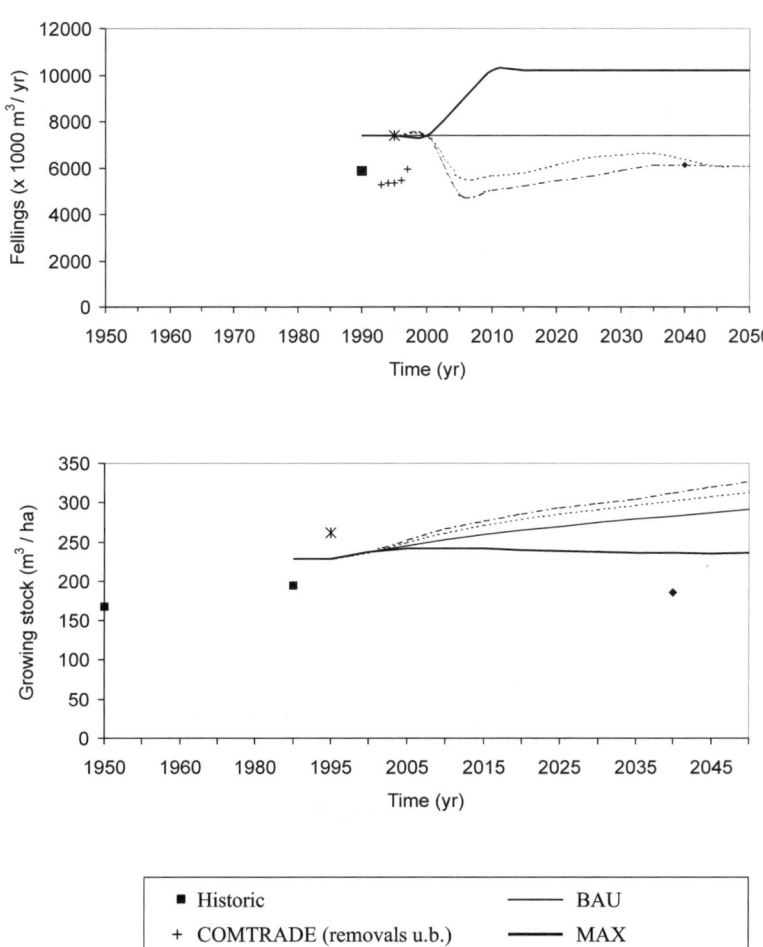

Fig. 5.65. Projected total national fellings (top) and growing stock per ha (bottom) in Slovak forests under the four scenarios until 2050.

The gross annual increment shows approximately the same trend for all scenarios; slightly declining until 2020, and then rather stable around 5.7 m^3/ha/yr in 2050.

The age class distributions show some ageing in the BAU scenario and in the MTF scenario. The MAX scenario leads to more area in

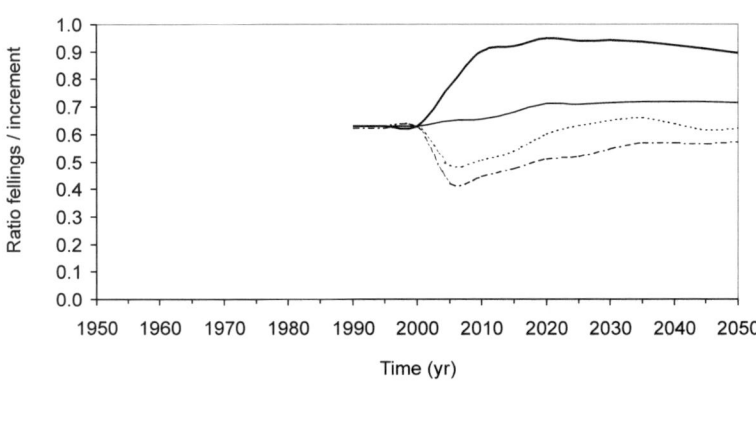

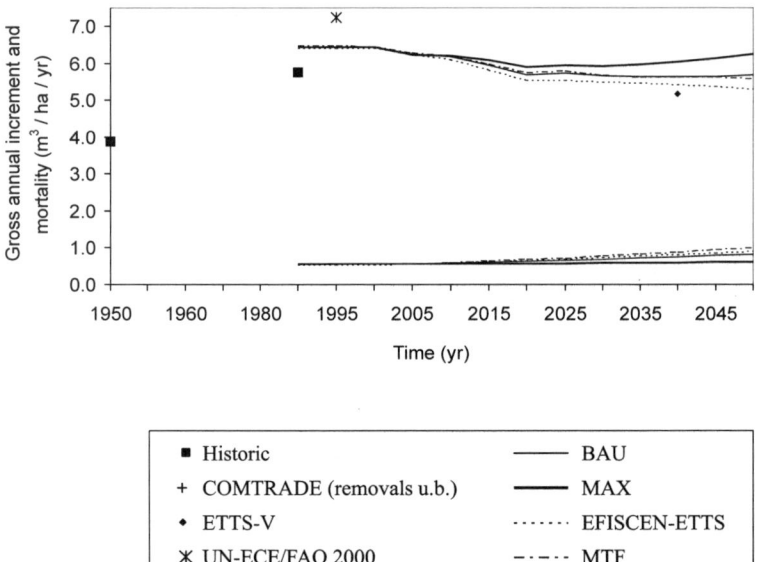

Fig. 5.66. Ratio of projected fellings/increment in Slovak forests under the four scenarios until 2050 (top) and projected net annual increment and mortality (bottom).

the younger age classes. This results in small regeneration areas, and thus small areas of forest in the young and middle age classes in 2050.

A total production of dead wood of 1.9 million m^3/yr is achieved in the MTF scenario. Due to the low utilisation level, there is ample

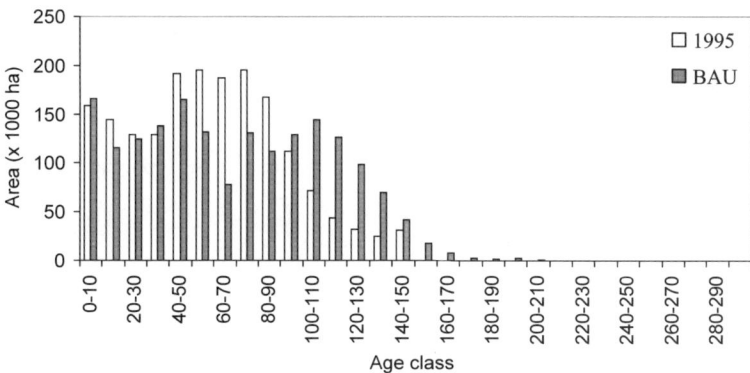

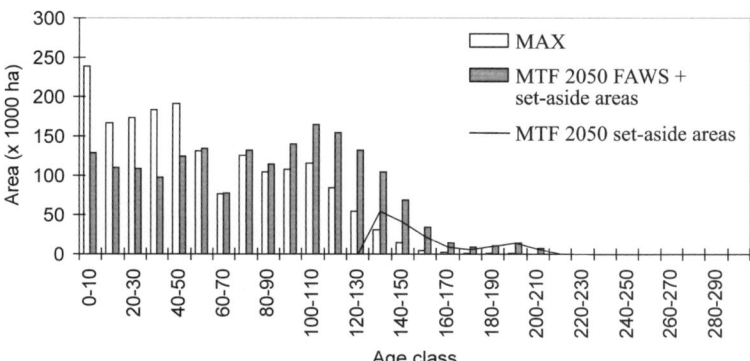

Fig. 5.67. The age class distribution of the Forest Available for Wood Supply (FAWS) in the initial situation and for 2050 under the BAU scenario (top). The age class distribution of FAWS for 2050 under the MAX scenario, the MTF scenario including set-aside areas and the MTF scenario with only set-aside areas (bottom).

possibility in Slovakian forests for nature-oriented management, having set aside 163 000 ha for reserves in 2050.

REFERENCES

Scheer, L. and Longauer, R. 1999. The Slovak Republic. In: Pelkonen, P., Pitkänen, A., Schmidt, P., Oesten, G., Piussi, P. and Rojas, E. (eds.), Forestry in Changing Societies in Europe. Part II. SILVA Network. University Press, Joensuu, Finland. Pp. 335–355.

UN-ECE/FAO 2000. Forest Resources of Europe, CIS, North America, Australia, Japan and New Zealand. Geneva, Timber and Forest Study Papers, No 17. 445 p.

Further information

Ministry of Agriculture 1997. Report of forestry in Slovak republic. Ministry of Agriculture, Bratislava. 123 p.

5.24. SLOVENIA

In co-operation with Dr. Hocevar

Introduction

With about half of the land area covered by forests, Slovenia is one of the more densely forested countries in Europe. The forest area is about 1.1 million ha of which two-thirds are privately owned. Much of the very diverse forests are in the mountainous regions where the Mediterranean climate is milder (Boncina and Winkler 1999). About two-thirds of the forests are broadleaved or mixed species forests, producing high quality timber. Some 10% of the forests are managed in a coppice system (UN-ECE/FAO 2000). The forest sector contributes 0.5% to the national GDP, but is very important in terms of protection values, and for local needs for fuelwood collection and tourism.

Country specific scenario assumptions

The data underlying the projections for Slovenia cover 1.07 million ha, which is 3% more than the UN-ECE/FAO (2000) figure for the forest area available for wood supply. In our underlying data, 50 000 ha of this area is under a coppice type management system (being less than the official statistics indicate). The data represent the inventory cycle of 1991–1995 and were distinguished by two owner classes and five tree species.

Business-as-usual scenario

The initial year was assumed to represent the situation in 1995. UN-ECE/FAO (2000) reports a felling of 2.3 million m^3 of roundwood per year of which 1.5 million m^3 is in coniferous forests. The COMTRADE removals data show a rapid increase in Slovenian removals underbark, from 1.8 million m^3/yr in 1993 to 2.2 million m^3/yr in 1997. The projections were initialised with fellings according to UN-ECE/FAO (2000): 2.3 million m^3/yr, of which 65% is from coniferous high forests, 30% from deciduous high forests, and 5% from coppice type management systems. For the high forests, 30% of total fellings was assumed to come from thinnings, and for coppice this was set at 15%. It was assumed that the area of forest will stay at the 1995 level and that there is no change in species composition.

EFISCEN-ETTS scenario

Although not distinguished separately at that time, the same trend in projected harvesting was assumed here for Slovenia as ETTS foresaw for the whole of Yugoslavia. Therefore, ETTS-V foresaw a sharp increase in fellings that would mean (relative to the proportion of forest area in Slovenia) an increase from 2.3 million m^3/yr in 1995 to 2.9 million m^3/yr in 2040; a 15% increase in coniferous high forests, 44% in deciduous high forests, and 50% in coppice. It was assumed that the area of forest will stay at the 1995 level. For the high forests, 30% of total fellings was assumed to come from thinnings, and for coppice this was set at 15%.

Maximum sustainable production scenario

The fellings were quickly increased to arrive at a point where fellings equal increment; i.e. a total level of 5.3 million m^3/yr by 2010. This was achieved by a doubling of the felling level in coniferous high forest, a tripling in deciduous high forest, and a small decrease in coppice.

Multi-functional scenario

The initial annual fellings of the BAU scenario was applied. It was assumed that the fellings increase of ETTS-V may be realistic, and the pattern was followed in this scenario (an increase from 2.3 million m^3/yr in 1995 to 2.9 million m^3/yr in 2040; a 15% increase in coniferous high forests, 44% in deciduous high forests, and 50% in coppice). A forest area increase of 60 000 ha in coniferous high forest was assumed for the period 1995–2015. With the economic recovery supporting it, it was assumed that there will be a strong desire for nature-oriented forest management. The rotation lengths were prolonged by 20 years in high forest and by 10 years in coppice. The share of thinnings from total fellings gradually increased to 50% in coniferous high forest, 40% in deciduous high forest, and 20% in coppice. We set aside broadleaved high forests over 140 years of age and mixed high forests of over 140 years of age. Initially this is an area of 18 000 ha. Through ageing this can increase during the simulation.

Results for Slovenia

The most noteworthy results for Slovenian forests can be seen in the age class distributions for 2050. There, the result of a concentration in forests in the age classes 60–100 years in 1995 can be seen. A clear ageing in the high forest is visible under all scenarios, while a

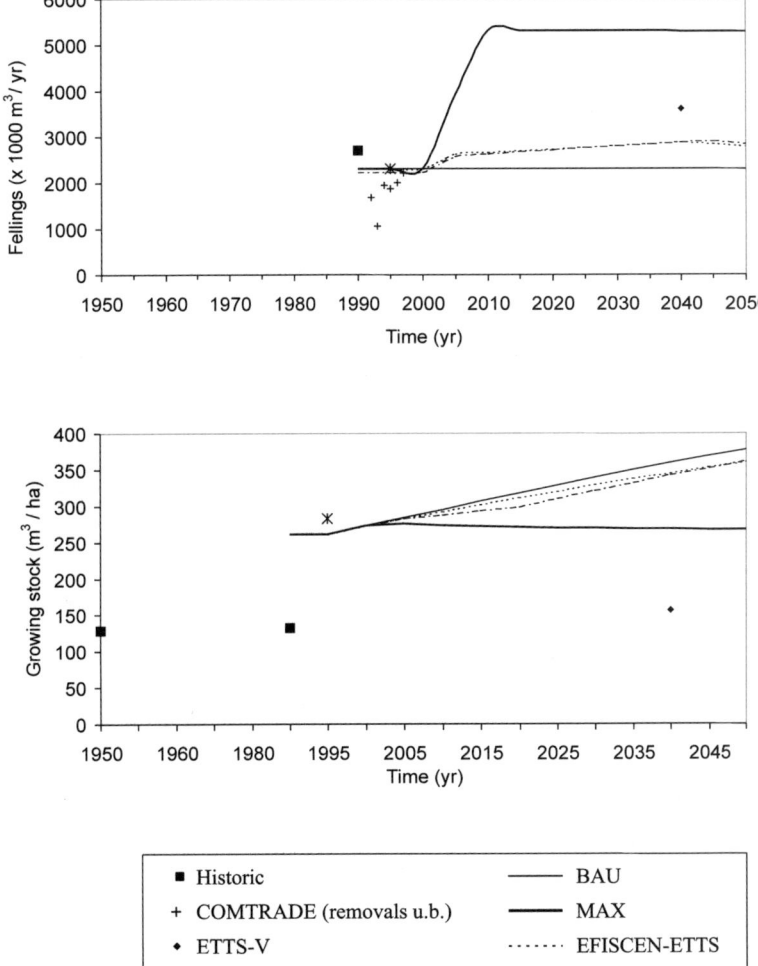

Fig. 5.68. Projected total national fellings (top) and growing stock per ha (bottom) in Slovenian forests under the four scenarios until 2050.

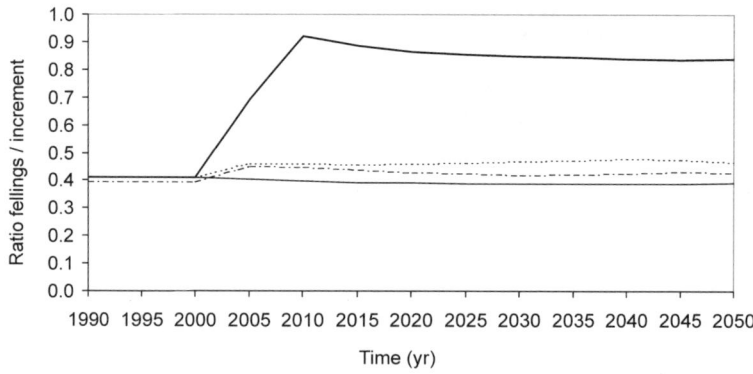

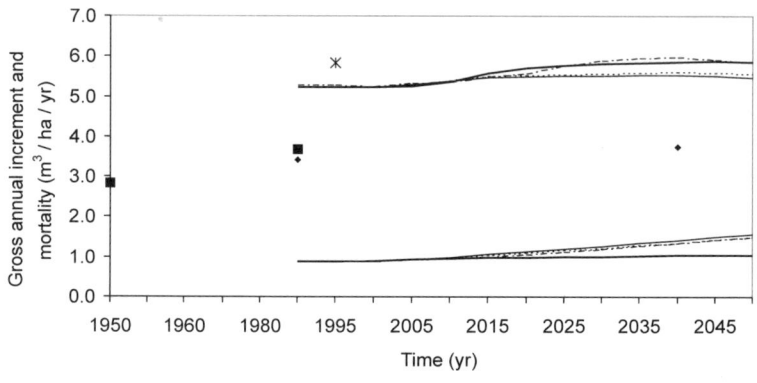

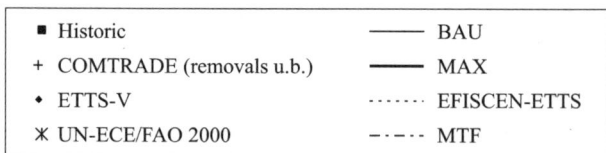

Fig. 5.69. Ratio of projected fellings/increment in Slovenian forests under the four scenarios until 2050 (top) and projected net annual increment and mortality (bottom).

small proportion of the forest was regenerated or is still under coppice management in 2050. Even the MAX scenario does not significantly change this.

The graphs show that only the latest international statistics (UN-ECE/FAO 2000) agree well with our initial values for growing stock

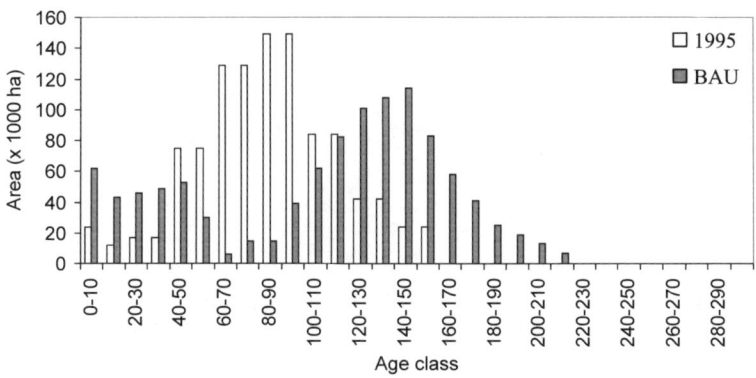

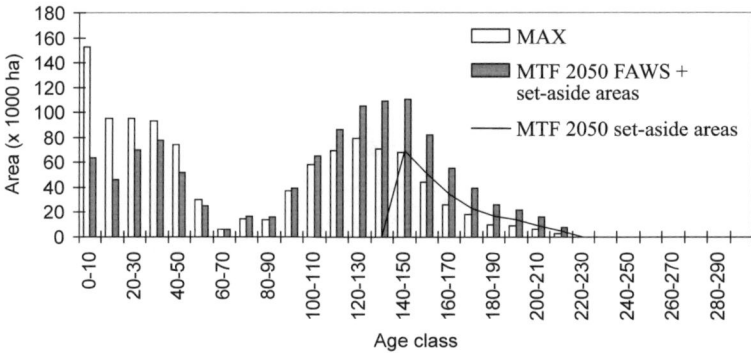

Fig. 5.70. The age class distribution of the Forest Available for Wood Supply (FAWS) in the initial situation and for 2050 under the BAU scenario (top). The age class distribution of FAWS for 2050 under the MAX scenario, the MTF scenario including set-aside areas and the MTF scenario with only set-aside areas (bottom).

and increment. The projections for Slovenia show that a maximum sustainable production of 5.3 million m^3/yr may be achieved. This is much higher than present levels. Under all other scenarios, the presently high average growing stock for Slovenia increases further.

Under the MTF scenario, the fellings gradually increase to 2.9 million m^3/yr in 2050, leading to an average growing stock of 361 m^3/ha by 2050. Under this scenario the utilisation level of the increment then still only amounts to 43%. In the MTF, mortality rates may then amount to 25% of the gross annual increment, producing 1.6 million m^3 of dead wood per year. The gross annual increment slightly increases in all scenarios from 5.2 m³/ha/yr initially, to 5.8 m³/ha/yr in 2050 due to an increase in the area of young forests in all scenarios.

In the MTF scenario 222 000 ha are projected to be newly established set-aside areas by 2050, not leading to problems in finding the required fellings.

REFERENCES

Boncina, A. and Winkler, I. 1999. Slovenia. In: Pelkonen, P., Pitkänen, A., Schmidt, P., Oesten, G., Piussi, P. and Rojas, E. (eds.), Forestry in Changing Societies in Europe. Part II. SILVA Network. University Press, Joensuu, Finland. Pp. 357–375.

UN-ECE/FAO 2000. Forest Resources of Europe, CIS, North America, Australia, Japan and New Zealand. Geneva, Timber and Forest Study Papers, No 17. 445 p.

5.25. SPAIN

In co-operation with Dr. J. Villanueva

Introduction

Most of the plateau that constitutes Spain has a dry Mediterranean climate. However, the mountain ranges to the north and northwest provide a wetter Atlantic climate. In the mountains, beech and fir dominate, while in the transition zone to the plateau, several species of pines dominate the forest. UN-ECE/FAO (2000) reports a present forest area of 13.5 million ha of which 9.2 million ha is in private ownership. The forest available for wood supply is 10.5 million ha (Madrigal et al. 1999). Of the wood produced, 81% comes from fast-growing species, while these occupy only 12% of the forest area. The forest sector contributes 2% to the GDP. Forest fires remain a threat to the Spanish forest resource.

Country specific scenario assumptions

The data underlying the projections covered 13.7 million ha of forest by diameter class distribution. Thus the uneven-aged modelling approach was used. The data represent 100% of the forest area according to UN-ECE/FAO (2000). The data are based on the 1986–1996 inventory cycle. The diameter class data were, in principle, distinguished by 50 regions and 4 to 46 species per region. However, the simulations distinguished 31 tree species at the national level.

Business-as-usual scenario

UN-ECE/FAO (2000) reports an annual felling of 12.7 million m^3 of roundwood per year, of which 7.3 million m^3 is from coniferous species. The COMTRADE removals data give a rather stable level of 15.5 million m^3/yr. ETTS-V, which was initialised also for 1990, starts with annual fellings of 15.5 million m^3/yr. Fellings were initialised with a utilisation level of 53% of the increment, i.e. 18.3 million m^3/yr. It was assumed that the area of forest will stay at the 1990s level and that there is no change in species composition.

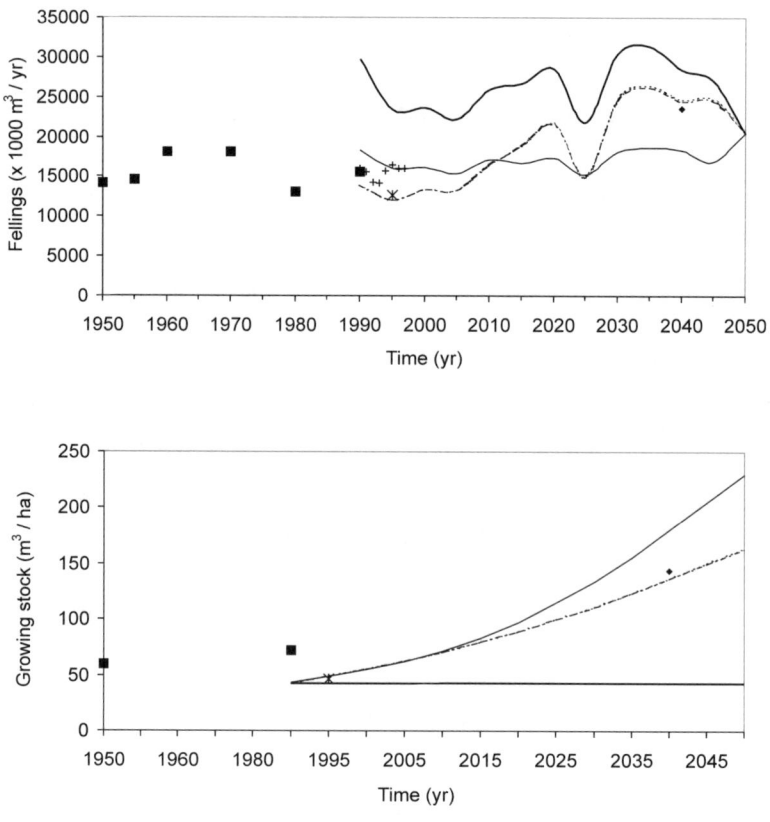

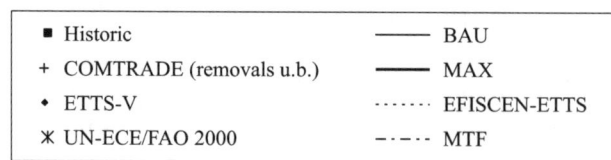

Fig. 5.71. Projected total national fellings (top) and growing stock per ha (bottom) in Spanish forests under the four scenarios until 2050.

EFISCEN-ETTS scenario

ETTS-V foresaw a strong increase in fellings in Spain from 15.5 million m³/yr in 1990 to 23.5 million m³/yr in 2040. Where possible, this was followed in this scenario. ETTS-V also foresaw an area

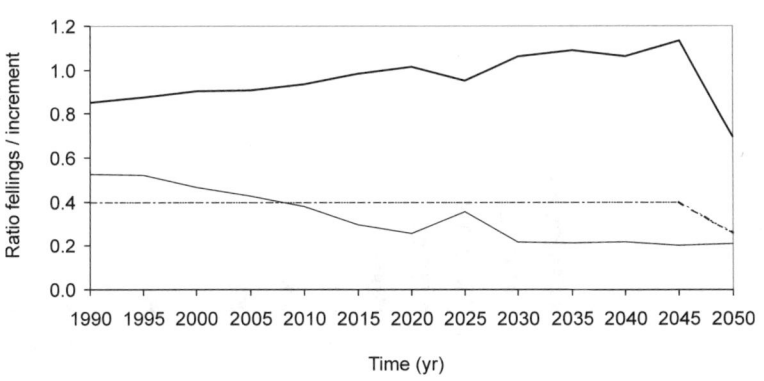

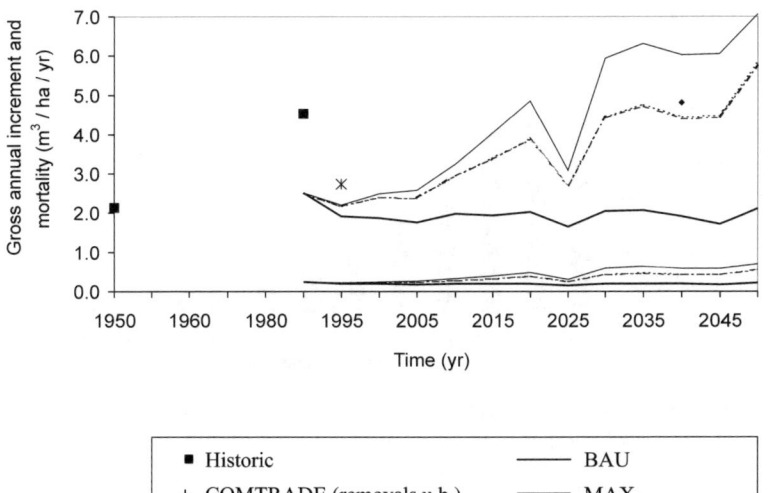

Fig. 5.72. Ratio of projected fellings/increment in Spanish forests under the four scenarios until 2050 (top) and projected net annual increment and mortality (bottom).

increase in the forest area available for wood supply, from 6.4 million ha in 1990 to 9.1 million ha in 2040. In the uneven-aged modelling approach it was not possible to follow this.

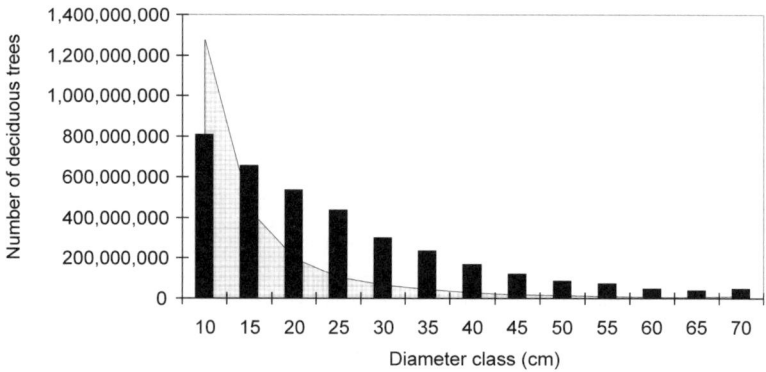

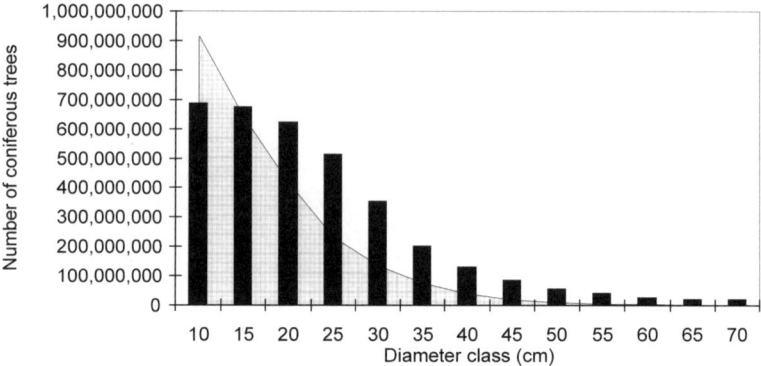

Fig. 5.73. Diameter class distributions for 1990 (area) and 2050 (bars) under the MTF scenario for deciduous trees (top) and for coniferous trees (bottom).

Maximum sustainable production scenario

The fellings were quickly increased to a point where fellings equal increment; i.e. on average a level of 26.2 million m^3/yr.

Multi-functional scenario

The felling increase of ETTS V was assumed to be realistic and was thus followed here. It decreases to a utilisation level of 40% of the increment. To take into account the current discussion concerning nature-oriented management, the area of six coniferous tree species initially over 40 cm in diameter and seven deciduous tree species initially over 45 cm in diameter were set aside. Through ageing this can increase during the simulation. No afforestation was assumed to take place.

Results for Spain

The different modelling approach required here, the limited options for controlling scenarios and the sensitivity of this approach to parameterising increment, mean that the results for Spain have to be regarded with care. However, the results do indicate a large potential for increasing fellings in Spain; double the present level of 12.7 million m^3 of roundwood per year. If the present level of fellings remains, the growing stock may rise to 229 m^3/ha in 2050, with natural mortality amounting to 10% of the gross annual increment.

Most scenarios (except the MAX) show sharp increases in growing stocks and increment. This is caused by the modelling approach that uses a relative increment for each diameter class. With shifts in the diameter classes as given in the figures above, increment and growing stock seem to grow exponentially. This approach works up to growing stock values of about 150 m^3/ha. Above this level it becomes very sensitive to the parameterised increment rates.

In the diameter class distributions for the MTF scenario, the ageing of the forest can clearly be seen. Through a set-aside scenario by diameter class, this leads to a sharp increase in area of set-aside through time: from 56 000 ha in 1990 to 1.0 million ha in 2050. Under the increment as projected, this does not hamper the fellings as foreseen under the MTF scenario.

REFERENCES

Madrigal, A., Fernandez-Cavada, J.L. and Ortuno, S.F. 1999. Spain. In: Pelkonen, P., Pitkänen, A., Schmidt, P., Oesten, G., Piussi, P. and Rojas, E. (eds.), Forestry in Changing Societies in Europe. Part II. SILVA Network. University Press, Joensuu, Finland. Pp. 377–400.
UN-ECE/FAO 2000. Forest Resources of Europe, CIS, North America, Australia, Japan and New Zealand. Geneva, Timber and Forest Study Papers, No 17. 445 p.

Further information

Dirección General de Conservación de la Naturaleza, 1996. Segundo Inventario Forestal Nacional 1986–1995. Ministerio de Medio Ambiente. Volumes per region.

5.26. SWEDEN

In co-operation with Prof. Ulf Söderberg

Introduction

The Swedish growing conditions vary from a temperate climate in the south to limited growing conditions in the boreal north. Sweden is the largest forest country in Europe with 27.2 million ha of forest (UN-ECE/FAO 2000). However, its total national wood increment (94 million m^3/yr) is in the same range as for France and Germany. Some 39% of the forest is owned by companies, 51% by private owners, and the rest is owned by the state (Lundkvist et al. 1999). The forest mainly consists of a few tree species: Norway spruce and Scots pine account for 46% and 38% of the growing stock, respectively. At present the forest sector contributes 20% to the total national export value. One of the recent challenges to the forest owners and managers is the move towards a more nature-oriented forest management.

Country specific scenario assumptions

The data underlying the projections covered 22.2 million ha of forest, while UN-ECE/FAO (2000) reported a forest area available for wood supply of 21.2 million ha. The data are based on the 1988-1992 inventory cycle. The data were distinguished by five regions, two owner classes, three site classes, and five species. Since broadleaves are usually mixed in other stands, the basic data underestimated the total growing stock for this species group. A correction to account for total growing stock of this group according to national statistics was carried out to take this into account.

Business-as-usual scenario

UN-ECE/FAO (2000) reported a production of 66 million m^3 of roundwood per year, of which 86% is from conifers. ETTS-V (Pajuoja 1995) used annual fellings of 62.4 million m^3/yr for 1990. The projections were thus initialised with fellings of 66 million m^3/yr. Some 40% of total fellings was assumed to come from thinnings. It

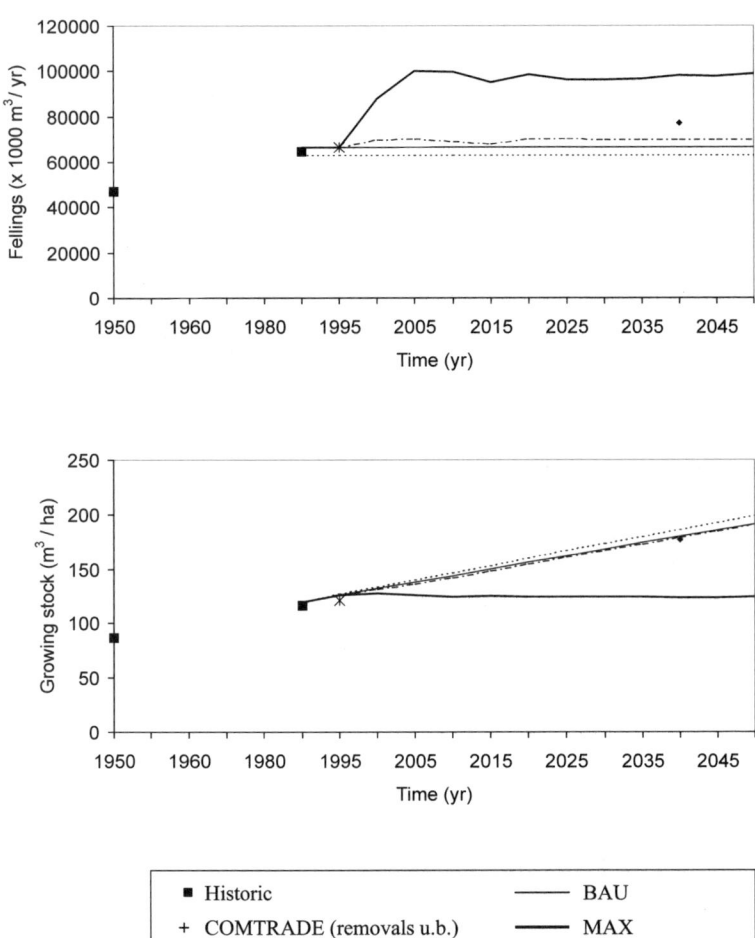

Fig. 5.74. Projected total national fellings (top) and growing stock per ha (bottom) in Swedish forests under the four scenarios until 2050.

was assumed that the area of forest will stay at the 1990s level and that there is no change in species composition.

EFISCEN-ETTS scenario

ETTS-V foresaw a very stable level of fellings until 2040 of 62.4 million m³/yr. ETTS-V foresaw no forest area expansion.

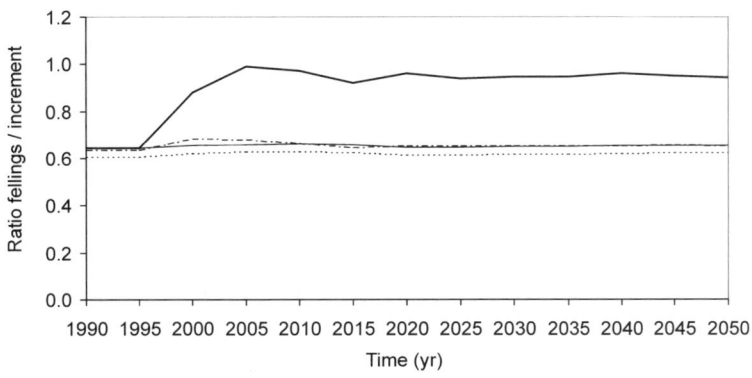
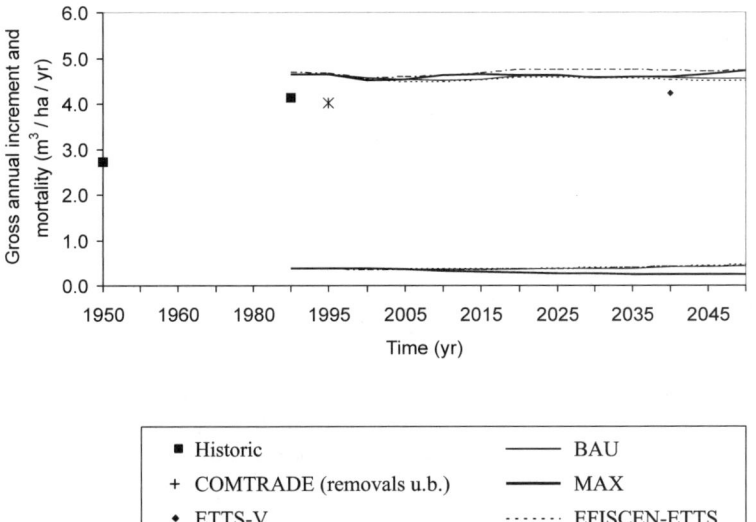

Fig. 5.75. Ratio of projected fellings/increment in Swedish forests under the four scenarios until 2050 (top) and projected net annual increment and mortality (bottom).

Maximum sustainable production scenario

The fellings were quickly increased to 103 million m³/yr in 2005 to a point where fellings equal increment. Relatively, the largest increase appeared possible in deciduous forests that by 2005 make up 18% of total fellings.

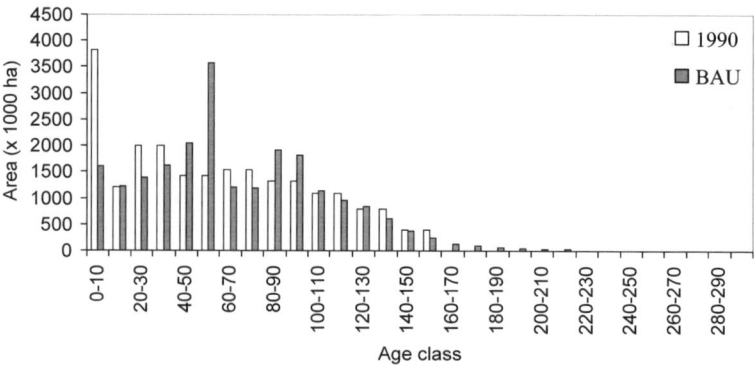

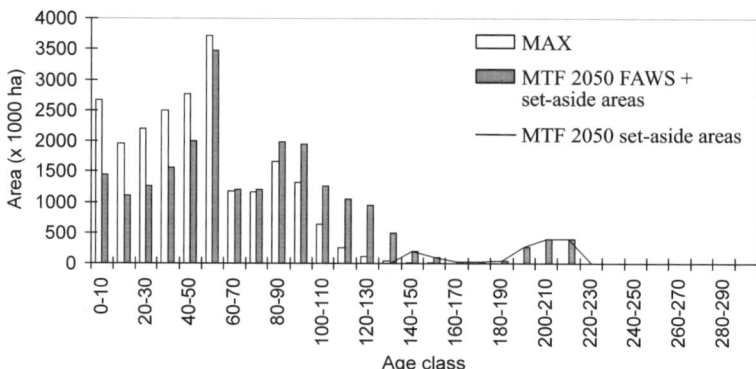

Fig. 5.76. The age class distribution of the Forest Available for Wood Supply (FAWS) in the initial situation and for 2050 under the BAU scenario (top). The age class distribution of FAWS for 2050 under the MAX scenario, the MTF scenario including set-aside areas and the MTF scenario with only set-aside areas (bottom).

Multi-functional scenario

Although the utilisation rate in Swedish forests is around 60% of the increment only, it was still assumed that Sweden will not increase its production much more than today. This is because of the move towards nature-oriented forest management. Therefore, we assumed that fellings will only increase by 0.8% per year for the first 5 years. After that they stabilise. All rotation lengths were prolonged by 20 years. The share from thinnings were increased to 50% from total fellings. All forests initially over 140 years of age were set aside as reserves. Initially this was an area of 808 000 ha. Through ageing this can increase during the simulation. A forest expansion of 209 000 ha between 1990 and 2010 was assumed.

Results for Sweden

The projections for the Swedish forests show that a maximum sustainable production of 97 million m^3/yr may be achieved, which is a large increase from the present level. Under this MAX scenario, the growing stock stays at 124 m^3/ha and the increment remains stable at 4.7 m^3/ha/yr. Under the MAX scenario, the age classes are pushed back towards younger forests.

Overall the projections show a rather inert forest resource, no large fluctuations and no large changes in increment nor in age class distributions. There are no large discrepancies with international statistics. The gross annual increment shows approximately the same trend for all scenarios; very stable around 4.6 m^3/ha/yr by 2050.

By 2050 the growing stock under the MTF scenario has grown to 190 m^3/ha. Under this scenario, mortality rates may then amount to 10% of the gross annual increment.

The projections for the Swedish forests age class distributions show only marked ageing in the MTF scenario where the set-aside policy starts to take effect by 2050; 1.48 million ha of reserves are established by this time. The moderate increase in fellings that we foresaw under this MTF scenario can, in that case, easily be found. A total production of dead wood of 10.2 million m^3/yr is achieved in the MTF scenario.

REFERENCES

Lundkvist, H., Lönnstedt, L., Persson, H. and Bäckström, P.-O. 1999. Sweden. In: Pelkonen, P., Pitkänen, A., Schmidt, P., Oesten, G., Piussi, P. and Rojas, E. (eds.), Forestry in Changing Societies in Europe. Part II. SILVA Network. University Press, Joensuu, Finland. Pp. 401–414.
Pajuoja, H. 1995. The outlook for the European Forest Resources and roundwood supply. Geneva Timber and Forest Discussion Papers ECE/TIM/DP/4. UN-ECE/FAO. Geneva. 59 p.
UN-ECE/FAO 2000. Forest Resources of Europe, CIS, North America, Australia, Japan and New Zealand. Geneva, Timber and Forest Study Papers, No 17. 445 p.

Further information

Lundström, A. and Söderberg, U. 1996. Outline of the HUGIN system for longterm forecasts of timber yields and possible cuts. In: Päivinen, R., Roihuvuo, L. and Siitonen, M. (eds.), Large scale forestry scenario models: experiences and requirements. EFI Proceedings 5. European Forest Institute. Pp. 63–78.

5.27. SWITZERLAND

In co-operation with Dr. Peter Brassel and D.I. Urs-Beat Brändli

Introduction

The topography of Switzerland varies considerably between the different regions in the country thus providing a wide diversity in ecological conditions. Much of the Swiss forests are situated at medium elevation heights in the Alps. Of the 1.17 million ha of forests, almost 70% is in public ownership, but mostly as communal forests (UN-ECE/FAO 2000). Norway spruce and fir constitute 50% of the forest resource (Schmithüsen and Zimmermann 1999). The good growing conditions and a long tradition of forestry have resulted in a high average growing stock (currently 336 m^3/ha). There is a clear trend towards a more nature-oriented management in Switzerland. This is in accordance with the protection function of much of the forest. The forest sector contributes 1.5% to the GDP.

Country specific scenario assumptions

The data underlying these projections covered 1.04 million ha of forest, while UN-ECE/FAO (2000) reported a forest area available for wood supply of 1.06 million ha. The difference is the clearcut areas. To compensate for this, in all scenarios an afforestation of 20 000 ha is carried out in the initial year. The data are based on the 1982–1986 inventory (even though at present a newer inventory is available). The data were distinguished by five regions, six site classes, and two tree species groups (coniferous and deciduous).

Business-as-usual scenario

UN-ECE/FAO (2000) reported a production of 7.4 million m^3 of roundwood per year, of which 5.4 million m^3/yr is from conifers. The COMTRADE removals underbark data show a rather stable level of fellings around 4.6 million m^3/yr. The projections were initialised according to UN-ECE/FAO (2000). Some 40% of total fellings was assumed to come from thinnings. It was assumed that the area of forest will stay at the 1985 level and that there is no change in species composition.

EFISCEN-ETTS scenario

ETTS-V first foresaw a decline in the felling compared with the storm year 1990, and then foresaw an increasing trend in fellings between 2000 and 2030. According to ETTS-V, fellings would total 6.7 million m^3/yr in 2040. ETTS-V foresaw a forest area expansion of 4000 ha by 2000.

Maximum sustainable production scenario

Fellings in both coniferous and deciduous forests were increased to a total of 11 million m^3/yr in 2005 to arrive at a point where fellings equal increment.

Multi-functional scenario

It was assumed that the Swiss will curb the increasing trend in growing stock, and that will be achieved by increasing fellings (supported by the wood market) with 1% per year until 2035. Therefore, by 2035 we arrive at a total requested felling level of 10.7 million m^3/yr. To represent the desire for nature-oriented forest management, all rotation lengths were prolonged by 20 years and the share of thinnings from total fellings stays at 40%. All coniferous forests initially over 160 years of age and deciduous forests initially over 140 years of age were set aside as reserves. Initially this represents 122 000 ha. Forest area expansion was assumed not to take place.

Results for Switzerland

Swiss results are affected by a wide range of assumptions on future fellings. The results show a large biological potential for increasing the fellings. In the simulated forests available for wood supply, the MAX scenario shows that a production of 9.9 million m^3/yr may be achieved. Under this felling level, the growing stock remains at 350 m^3/ha and increment at the high level of 10.2 m^3/ha/yr. All previous international statistics showed far lower increment estimates, thus hampering comparisons with our projections.

Under the MTF scenario (with the assumed wish to curb the growing stock increase) the increase of growing stock continues until 2020 to peak at 402 m^3/ha, after that it declines to 385 m^3/ha

SWITZERLAND

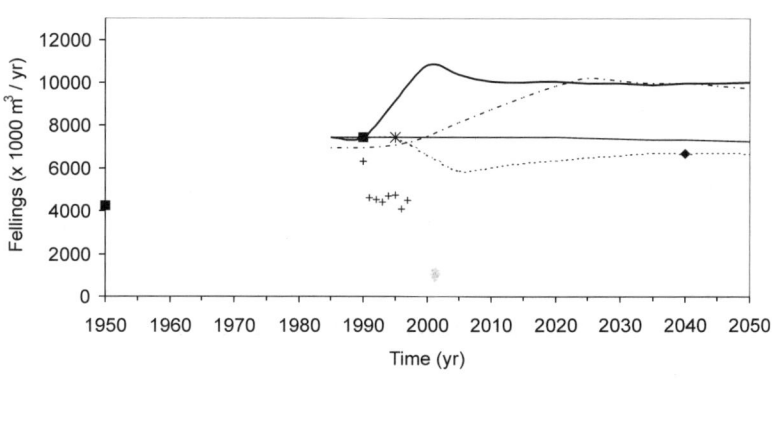

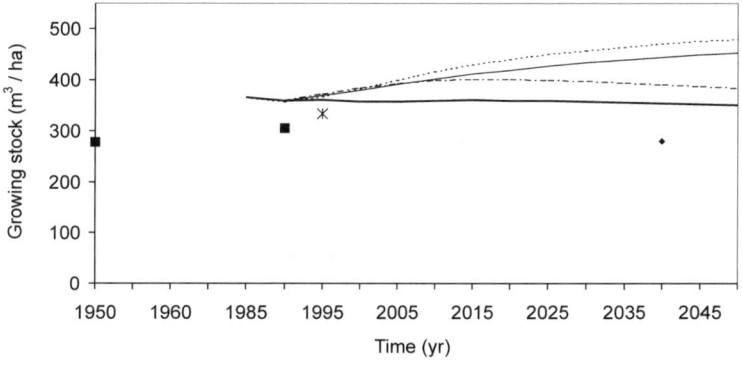

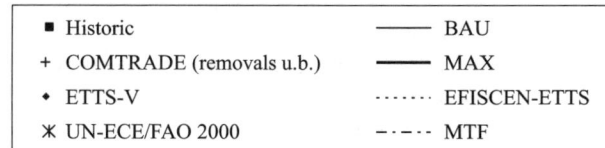

Fig. 5.77. Projected total national fellings (top) and growing stock per ha (bottom) in Swiss forests under the four scenarios until 2050.

in 2050. Mortality rates may then amount to 17% of the gross annual increment, or a total production of dead wood of 1.8 million m^3/yr. ETTS-V, which foresaw a much more moderate gross annual increment, consequently projected a much lower growing stock by 2050: 280 m^3/ha. Under the BAU scenario, the growing stock build-up continues and reaches 453 m^3/ha in 2050.

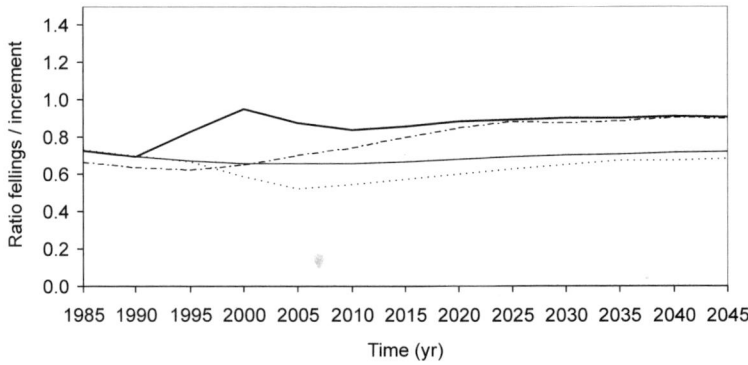

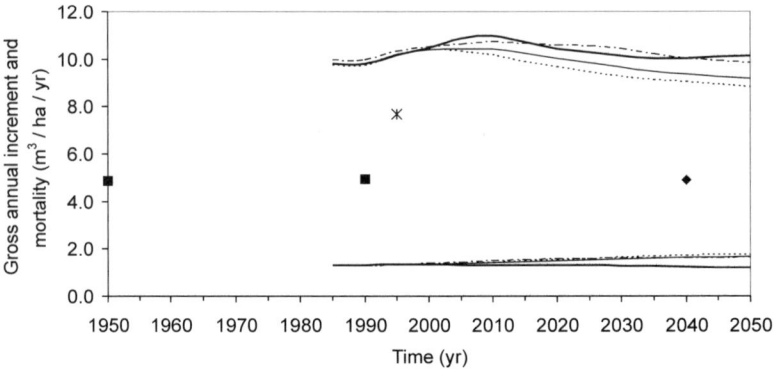

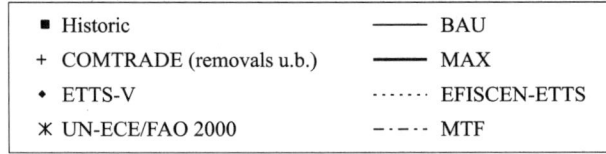

Fig. 5.78. Ratio of projected fellings/increment in Swiss forests under the four scenarios until 2050 (top) and projected net annual increment and mortality (bottom).

The gross annual increment shows approximately the same trend for all scenarios; slightly increasing until 2015, and then declining to values around 9.5 m³/ha/yr in 2050.

The age class distributions show marked ageing in the BAU scenario and in the MTF scenario (note that the small peak in age classes 160–

SWITZERLAND

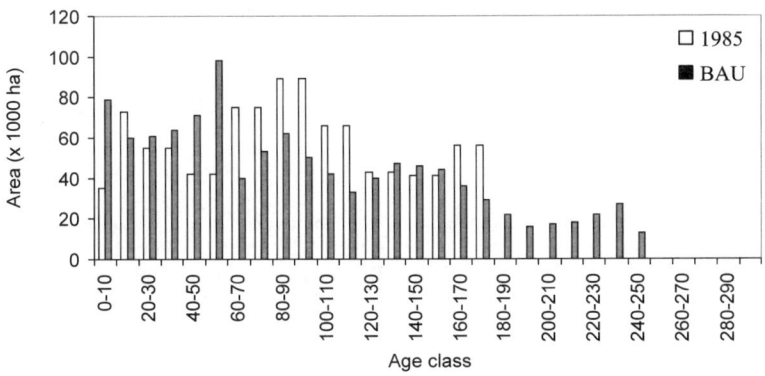

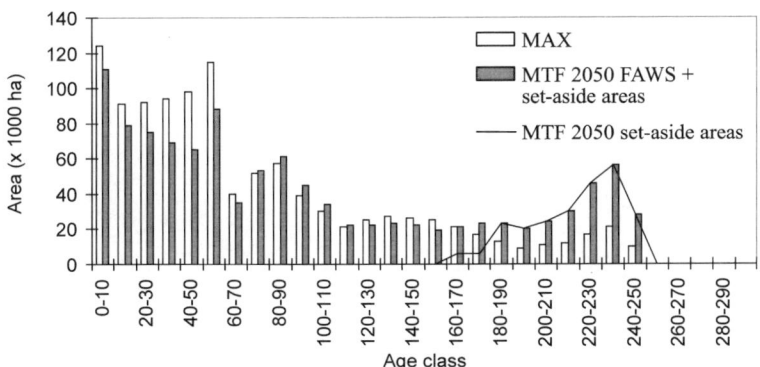

Fig. 5.79. The age class distribution of the Forest Available for Wood Supply (FAWS) in the initial situation and for 2050 under the BAU scenario (top). The age class distribution of FAWS for 2050 under the MAX scenario, the MTF scenario including set-aside areas and the MTF scenario with only set-aside areas (bottom).

180 years in the initial year 1985 are caused by grouping all forest over 160 years in these two classes). The MAX scenario leads to larger areas in the younger age classes. The assumed strong increase in fellings in the MTF scenario is not hampered by the 283 000 ha of reserves in 2050.

REFERENCES

Schmithüsen, F. and Zimmermann, W. 1999. Switzerland. In: Pelkonen, P., Pitkänen, A., Schmidt, P., Oesten, G., Piussi, P. and Rojas, E. (eds.), Forestry in Changing

Societies in Europe. Part II. SILVA Network. University Press, Joensuu, Finland. Pp. 415–441.

UN-ECE/FAO 2000. Forest Resources of Europe, CIS, North America, Australia, Japan and New Zealand. Geneva, Timber and Forest Study Papers, No 17. 445 p.

Further information

Brassel, P. and Brändli, U.-B. (eds.) 1999. Schweizerisches landesforstinventar, Ergebnisse der Zweitaufnahme 1993–1995. Eidgenössische Forschungsanstalt für Wald, Schnee und Landschaft, WSL. Birmensdorf. Verlag Paul Haupt. 442 p.

5.28. TURKEY

In co-operation with Mr. Ulvi Us and Ms. D. Berghmans

Introduction

Turkey consists primarily of an undulating plateau rising eastward to an altitude of some 2000 m. The plateau is bordered by high mountain ranges. Except for the coastal areas, Turkey has a continental climate with cold winters and hot summers. The total forest and other wooded land area amounts to 20.7 million ha, with the forest area amounting to 9.9 million ha. All of the forest area is publicly owned (UN-ECE/FAO 2000). The species composition of the forests in Turkey is very diverse. Due to water limitation, the growth of the forest is limited and the forests are under serious threat from wildfires. Turkey has major afforestation schemes. The forests' main function is protection and conservation.

Country specific scenario assumptions

The projections for Turkey have to be regarded with care, because they only cover 5.5 million ha out of 8.6 million ha of forest available for wood supply (UN-ECE/FAO 2000). Note that the FRA 1990 gave an area of exploitable forests of 6.6 million ha. The reason for the underestimation in our covered forest area is that not all regions of Turkey were included. We have mostly covered the northern, western and southwestern regions of the country, which are the most densely forested regions. The data are based on the 1990 situation and did not include the coppice systems. The data were distinguished by 20 regions and 15 tree species. The mixed uneven-aged forests of Turkey were represented in the even-aged age class distributions as an area in the oldest age class.

Business-as-usual scenario

To initialise the required national production, several sources were consulted. UN-ECE/FAO (2000) reports a production of 17.6 million m^3 of roundwood per year, of which 56% is from conifers. ETTS-V starts with annual fellings of 17.1 million m^3/yr on the 6.6 million ha of exploitable forest that they covered. The COMTRADE removals data provide a slightly increasing trend from 15.7 million m^3/yr u.b. in 1990 to 18 million m^3/yr in 1997. The projections were initialised with fellings of 14.2 million m^3/yr, of which 8.3 million m^3/yr is from conifers. Some 30% of total fellings from conifers was assumed to come from thinnings, and for deciduous species thinnings were assumed to account for 25% of fellings. It was assumed that the area of forest will stay at the 1990s level and that there is no change in species composition.

EFISCEN-ETTS scenario

ETTS-V foresaw a very stable level of fellings until 2040; 17.1 million m^3/yr. ETTS-V foresaw no forest area expansion.

Maximum sustainable production scenario

The fellings were increased in coniferous forests to a point where fellings equal increment; but they had to be decreased in deciduous forests; i.e. a total level of 13.6 million m^3/yr, of which 9.6 million m^3/yr is from coniferous forests.

Multi-functional scenario

The initial annual fellings of the BAU scenario was applied, and then it was assumed that the fellings in coniferous forests will increase by 1% per year for the next 15 years. After that fellings in conifers stabilise. The fellings in deciduous forests were decreased following the outcome of the MAX scenario. The rotation lengths were prolonged by 20 years. The share from thinnings was increased to 50% from total fellings. The area initially over 130 years of fir, cedar, mixed conifer, beech, conifer and deciduous, other deciduous, oak, juniper and spruce were set aside as reserves. The area initially over 65 years of age of other conifer and alder were also set aside as reserves. Initially this represents an area of 196 000 ha. Through ageing this can increase during the simulation. Afforestation was assumed to take place at a slower pace than foreseen in the policy plans, i.e. 60 000 ha in total.

Results for Turkey

Results for Turkey may indicate that the present fellings are at their biological maximum, or slightly above. However, this result should be viewed with care as we have covered only part of the forests available

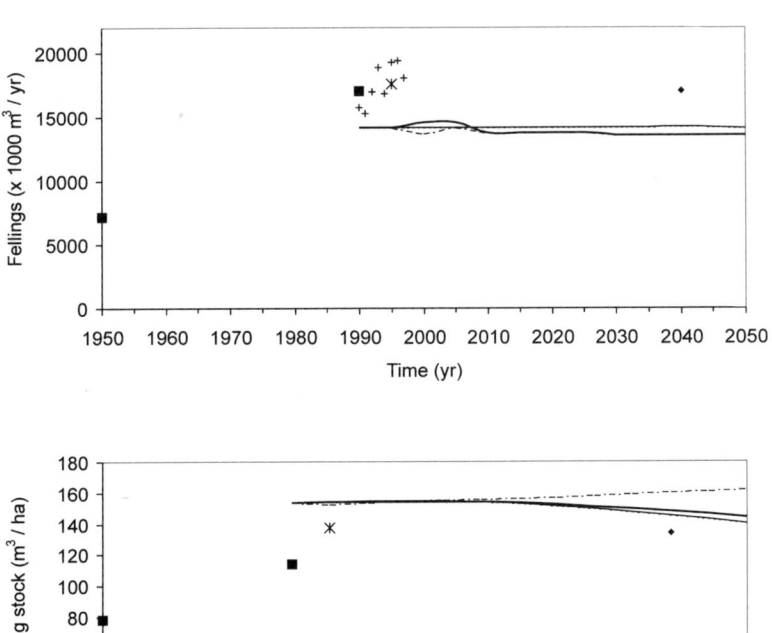

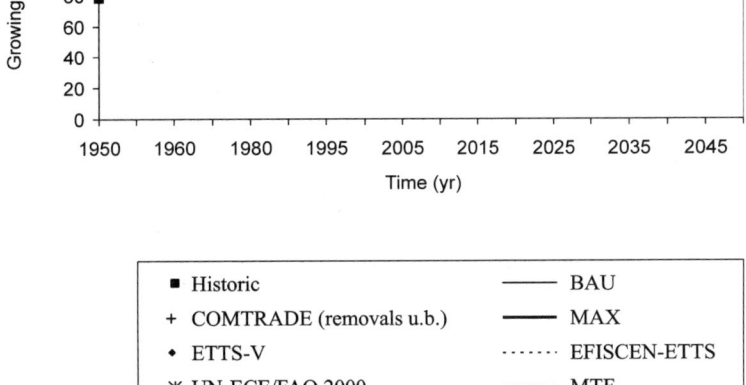

Fig. 5.80. Projected total national fellings (top) and growing stock per ha (bottom) in Turkish forests under the four scenarios until 2050.

for wood supply (although fellings in our scenarios were adapted to take this into account), and the increment data seemed to be quite uncertain.

Taking all this into account, the projections indicate that under the present fellings level, the average growing stock may decline from 153 m^3/ha at present to 140 m^3/ha in 2050. The projections for the 5.5

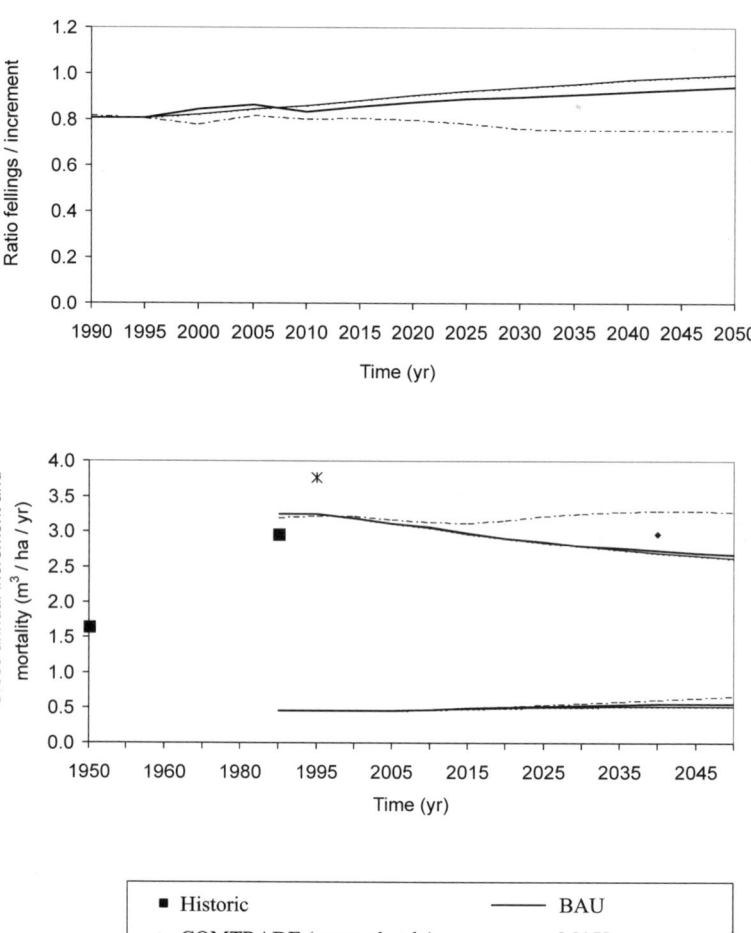

Fig. 5.81. Ratio of projected fellings/increment in Turkish forests under the four scenarios until 2050 (top) and projected net annual increment and mortality (bottom).

million ha of forests show that a maximum sustainable production of 13.6 million m³/yr may be achieved, which is a moderate decrease from the present level of around 14.2 million m³/yr. By 2050 the growing stock under the MTF scenario may amount to 162 m³/ha.

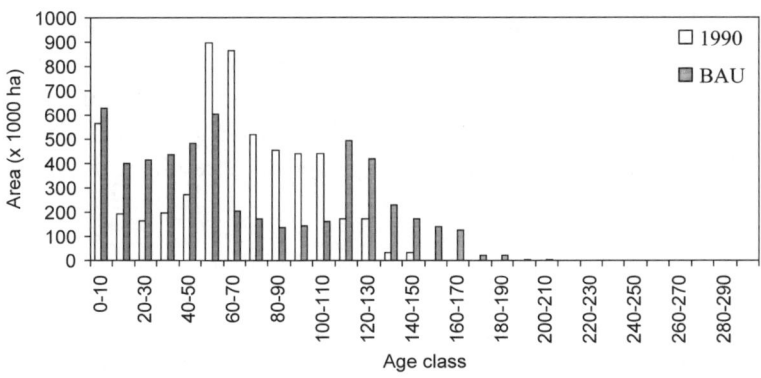

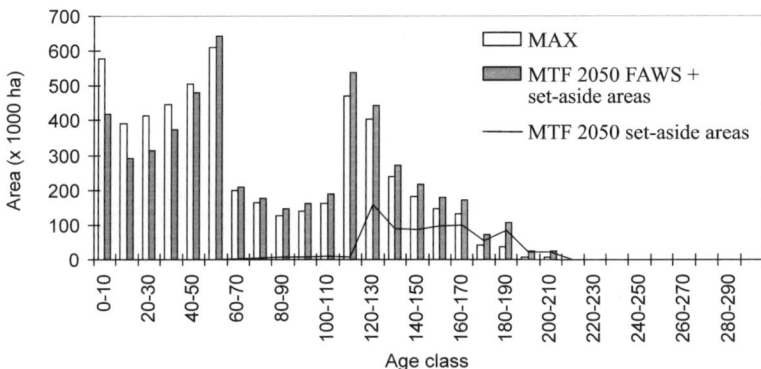

Fig. 5.82. The age class distribution of the Forest Available for Wood Supply (FAWS) in the initial situation and for 2050 under the BAU scenario (top). The age class distribution of FAWS for 2050 under the MAX scenario, the MTF scenario including set-aside areas and the MTF scenario with only set-aside areas (bottom).

The 1990 age class distribution for Turkish forests shows a concentration in the middle to older age classes. This is partly because the uneven-aged forest were represented in age through the oldest age class only. Some rejuvenation of the forest occurs in all scenarios. The MTF scenario shows the strongest ageing, because the set-aside policy is assumed to work out intensively in Turkey. By 2050, an area of 753 000 ha is projected to be newly established set-aside areas in the MTF scenario. This leads to problems in finding the required fellings: the required 1% increase in fellings under the MTF was not found. Instead, fellings decrease by a total of 5% between 1990 and 2030.

The MTF scenario maintains the highest increment (3.2 m³/ha/yr), mainly because more thinnings are applied in this scenario. Under this latter scenario, mortality rates may then amount to 19% of the gross annual increment, or 3.5 million m³ dead wood per year. This high mortality mainly occurs due to ageing, and as a result of the set-aside policy.

REFERENCES

UN-ECE/FAO 2000. Forest Resources of Europe, CIS, North America, Australia, Japan and New Zealand. Geneva, Timber and Forest Study Papers, No 17. 445 p.

For further information

Konukcu, M. 1998. Statistical profile of Turkish Forestry. Ankara, State Planning Organization.

5.29. UNITED KINGDOM

In co-operation with Simon Gillam

Introduction

With 10% of the land covered by forest, the UK is one of the less densely forested countries in Europe. Of the 2.4 million ha of forest, 1.3 million ha is privately owned (UN-ECE/FAO 2000). Fast-growing conifer plantations (largely in the upland regions and consisting of Sitka spruce) account for 87% of the total national fellings. There has been a 30% increase in forest area in the UK since 1950, mainly as a result of afforestation with this fast-growing spruce species. The large-scale afforestations provide a young forest estate now (Price and Samuel 1999), and have led to discussion on their ecological sustainability and landscape value. The forestry sector contributes 0.5% to the GDP, and the UK nowadays has a forest management policy aiming at more natural forests.

Country specific scenario assumptions

UN-ECE/FAO (2000) presents a forest area available for wood supply in the United Kingdom of 2.1 million ha. The data underlying the projections here cover 1.9 million ha. The under-representation is caused by the fact that these data are based on the 1979–1982 census and represent the exploitable forest area of that time. A new inventory is becoming available, but was not used here. The data were distinguished by four regions and seven tree species.

Business-as-usual scenario

UN-ECE/FAO (2000) reports a production of 9.5 million m^3 of roundwood per year, of which 8.3 million m^3/yr is from coniferous species. ETTS-V started with annual fellings of 6.4 million m^3/yr. The COMTRADE removals u.b. data show a steady increase in the United Kingdom, from 6.4 million m^3/yr in 1990 to 7.5 million m^3/yr in 1997. The projections were initialised according to the UN-ECE/FAO (2000) figures, taking into account that we cover 1.9 million ha only: 1.92 million m^3/yr. Some 33% of total fellings was assumed to come from thinnings. It was assumed that the area of forest will stay at the 1990s level with no species change.

EFISCEN-ETTS scenario

ETTS-V predicted a very sharp increase in fellings in the United Kingdom, from 6.4 million m^3/yr in 1990 to 16.8 million m^3/yr in 2040. An attempt was made to follow this pattern. ETTS-V predicted a large forest area expansion of 750 000 ha between 1990 and 2040; that was also followed here, allocating this new area of forests over all species, but mostly to Scots pine.

Maximum sustainable production scenario

The fellings (on the initial area) were quickly increased to a point where fellings equal increment; i.e. around 14 million m^3/yr by 2010. However, it was immediately seen that this level fluctuates strongly in the United Kingdom due to its young and skewed age class distribution.

Multi-functional scenario

The initial annual fellings of the BAU scenario was applied. Then it was assumed that the fellings in deciduous forests will increase by some 2–3% per 5 years until the end of the simulation period. For coniferous species it was assumed that fellings will increase by some 5% per 5 years until 2040. By 2050, fellings have thus reached a level of 11.5 million m^3/yr. It was assumed that the ETTS-V forest expansion scenario is close to feasible. To take into account the desire for a more nature-oriented forest management, the rotation lengths were prolonged by 10 years. The share of thinnings from total fellings gradually increased to 45%, and more than half of the afforestation is now done with Scots pine and deciduous species. We set aside Scots pine forest of more than 90 years, oak of more than 90 years and beech forests of more than 120 years of age. In this way 102 000 ha was set aside initially. Through ageing this can increase during the simulation.

Results for the UK

The results for the UK forests first of all show a strongly fluctuating nature of the fellings in the MAX scenario. This strongly fluctuating level is caused by the skewed and young age class distribution initially.

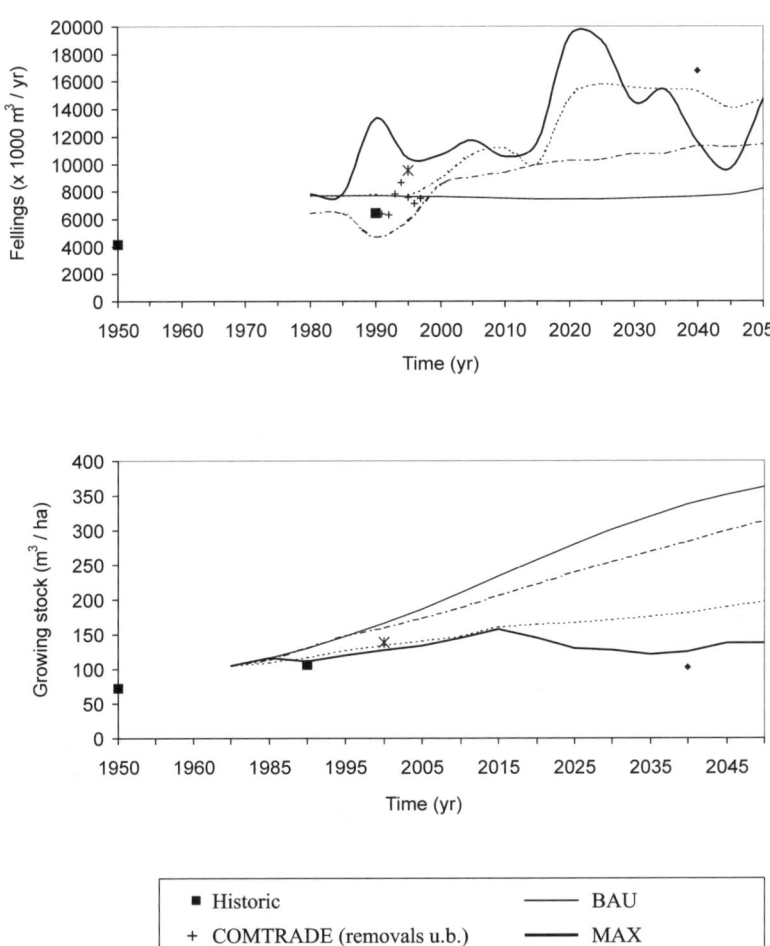

Fig. 5.83. Projected total national fellings (top) and growing stock per ha (bottom) in the UK forests under the four scenarios until 2050.

Therefore, from one time step to another there are large areas of forest that reach an age at which they can be felled. This occurs in the period 2020–2025, when the maximum felling level is assessed at just over 19 million m^3/yr. The long-term average sustainable felling level is, however, much lower at 12.5 million m^3/yr. The strongly fluctuating felling level in the MAX scenario also leads to

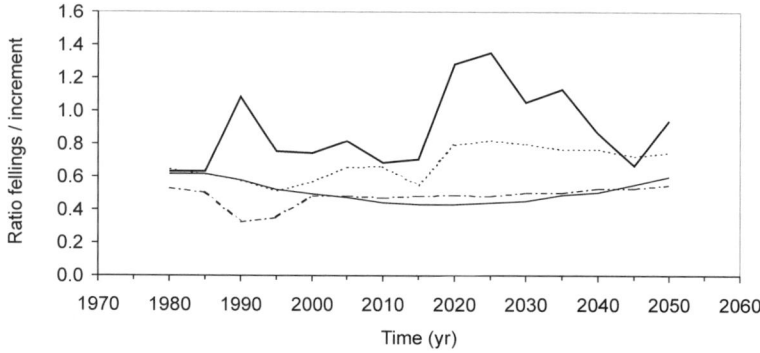

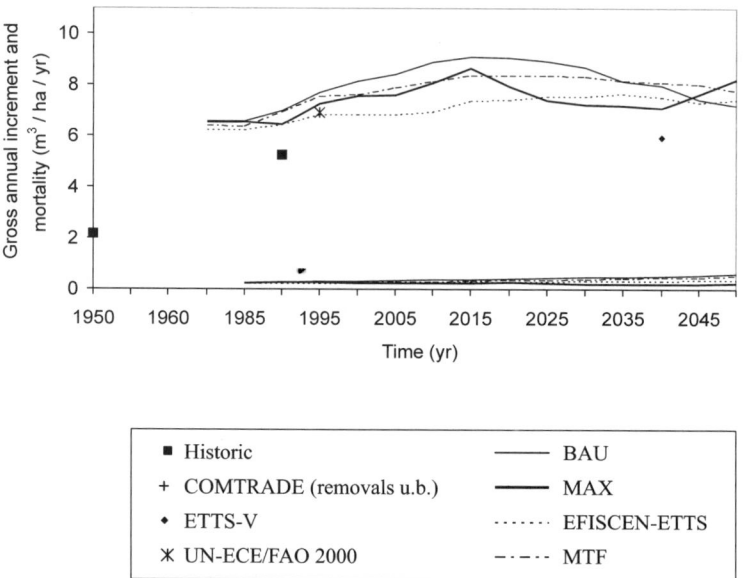

Fig. 5.84. Ratio of projected fellings/increment in British forests under the four scenarios until 2050 (top) and projected net annual increment and mortality (bottom).

a fluctuating felling/increment ratio, fluctuating volume development and a fluctuating increment (around 7.4 m^3/ha/yr).

Both the EFISCEN-ETTS and the MTF scenarios reflect the desire to increase forest area in the UK. In these scenarios the forest area was increased by 40% by 2050. Under the EFISCEN-ETTS the desired

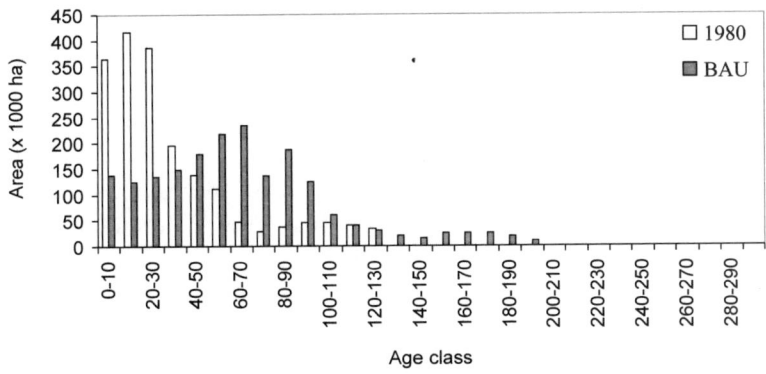

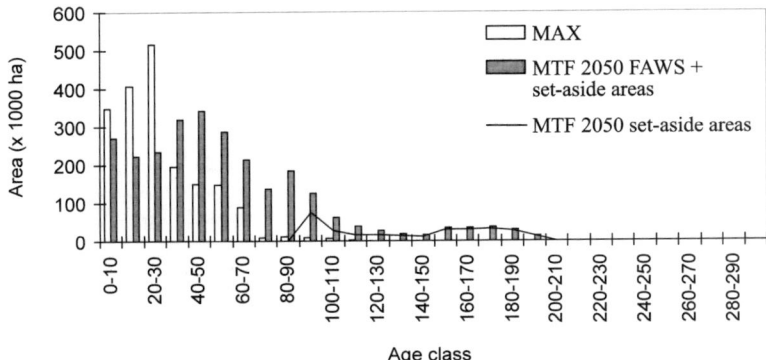

Fig. 5.85. The age class distribution of the Forest Available for Wood Supply (FAWS) in the initial situation and for 2050 under the BAU scenario (top). The age class distribution of FAWS for 2050 under the MAX scenario, the MTF scenario including set-aside areas and the MTF scenario with only set-aside areas (bottom).

increase in felling level cannot be found all the time. Around 2015, fellings stabilise and decline slightly (as can also be seen in the MAX scenario). In the MTF scenario the desired felling increase was much more gradual, and was achieved. The fellings stabilise at 11.3 million m^3/yr. However, the MTF scenario also shows that fellings cannot be found at the beginning of the simulation. This is due to the large area of reserves established initially (102 000 ha). The reserves area increases to 291 000 ha by 2050, but by that time this does not hamper the required fellings level.

The rather large-scale afforestations in EFISCEN-ETTS and the MTF scenarios show that their impact (new areas with no growing stock and low increment) on reducing national averages is modest in the UK. Therefore, the growing stock still increases considerably under the MTF scenario to 314 m^3/ha by 2050. Under the BAU scenario it may amount to 362 m^3/ha by 2050. Under MTF scenario, mortality rates may then amount to 7% of the gross annual increment; producing 1.4 million m^3 of dead wood per year.

The gross annual increment first increases in all scenarios from 6.4 m^3/ha/yr initially, to around 8.5 m^3/ha/yr in 2025, and then declines again to 7.7 m^3/ha/yr in 2050. This is clearly an effect of ageing. The age class distribution for 2050 clearly shows the strong ageing process under the BAU and the MTF scenarios because the utilisation ratio is around 45–50% only.

REFERENCES

Price, C. and Samuel, J. 1999. United Kingdom. In: Pelkonen, P., Pitkänen, A., Schmidt, P., Oesten, G., Piussi, P. and Rojas, E. (eds.), Forestry in Changing Societies in Europe. Part II. SILVA Network. University Press, Joensuu, Finland. Pp. 461–480.

UN-ECE/FAO 2000. Forest Resources of Europe, CIS, North America, Australia, Japan and New Zealand. Geneva, Timber and Forest Study Papers, No 17. 445 p.

Further information

Whiteman, A. 1991. The supply and demand for wood in the United Kingdom. Occasional paper 29. Forestry Commission, United Kingdom. 52 p.

Forestry Commission 1986. Forestry, facts and figures 1984/85. Forestry Commission. Alice Holt Lodge, Farnham, United Kingdom.

5.30. YUGOSLAVIA

Introduction

Yugoslavia officially comprises Serbia, Montenegro, Vojvodina and the UN protectorate of Kosovo. The southern half of this country is mountainous, while the north is formed by the Danube River plain. The climate varies from Mediterranean along the Adriatic Coast and in the south of the country, to continental and temperate further north and inland. About 28% of the land area is covered with forests (2.89 million ha). Most of the very diverse and mostly deciduous forests are situated in the mountainous south and southwest of the country. Some 53% of the forest is publicly owned, and much is under a coppice type management (UN-ECE/FAO 2000). The forests are important in terms of protection values and for local needs for fuelwood collection.

Country specific scenario assumptions

The data underlying the projections cover 1.51 million ha, which is only 65% of the figure UN-ECE/FAO (2000) gives for the forest area available for wood supply. Out of the covered area, 1.18 million ha is under coppice type management. The data cover only the regions Serbia and Montenegro and are still based on the data from Nilsson et al. (1992). Since it appeared impossible to update the data, it is unclear what areas are missing. The data represent the state of the forest around the early to mid-1980s, and were distinguished by two regions, two owner classes and four tree species.

Business-as-usual scenario

The initial year was assumed to represent the situation in 1985. UN-ECE/FAO (2000) reports a felling of 3.4 million m^3 of roundwood per year for 1991, of which 2.8 million m^3/yr is from deciduous forests. The COMTRADE removals data show a very stable level of removals underbark: 1.3 million m^3/yr. Taking into account the fact that we only cover 65% of the forest area available for wood supply, but assuming that the felling pressure is higher due to unreported fuelwood collection, we intialise fellings at a level of 3.1 million m^3/yr, of which 77% is from coppice type management systems. For the coniferous high forests, 30% of total fellings was assumed to come from thinnings, for deciduous high forests 45% of total fellings was assumed to come from thinnings, and for coppice the figure was 30%. It was assumed that the area of forest will stay at the 1985 level and that there is no change in species composition.

EFISCEN-ETTS scenario

Although not distinguished separately at that time, the same trend in projected harvesting was assumed here for Yugoslavia as ETTS foresaw for the larger Yugoslavia. ETTS-V foresaw a sharp increase in fellings between 1990 and 2000, and then a moderate increase of some 5% per year until 2040. Relative to the proportion of forest area covered here, an increase from 2.8 million m^3/yr in 1985 to 3.3 million m^3/yr in 2020 is foreseen. It was assumed that the area of forest will stay at the 1985 level.

Maximum sustainable production scenario

The fellings were quickly increased to arrive at a point where fellings equal increment; i.e. a total level of 4.2 million m^3/yr by 2010. This was achieved by lowering the felling level in coniferous high forest by 30%, a 30% increase in deciduous high forest, and a 45% increase in fellings in coppice systems.

Multi-functional scenario

The initial annual fellings of the BAU scenario was applied. Then it was assumed that the fellings increase of ETTS-V may be realistic, but we assumed that it may take longer before the economic recovery of Yugoslavia begins. Therefore, the fellings increase will only start by 2005–2010, when (in principle) it will amount to an increase of 7% per 5 years. This continues until 2020, and then levels off. The increase in fellings will mainly take place in the coppice system. However, at the same time, it was assumed that the population will not depend on fuelwood collection anymore, and that the coppice system is abandoned and converts to high forest. It was assumed that the economic recovery will also result in forest area expansion programmes, leading to a forest area increase of

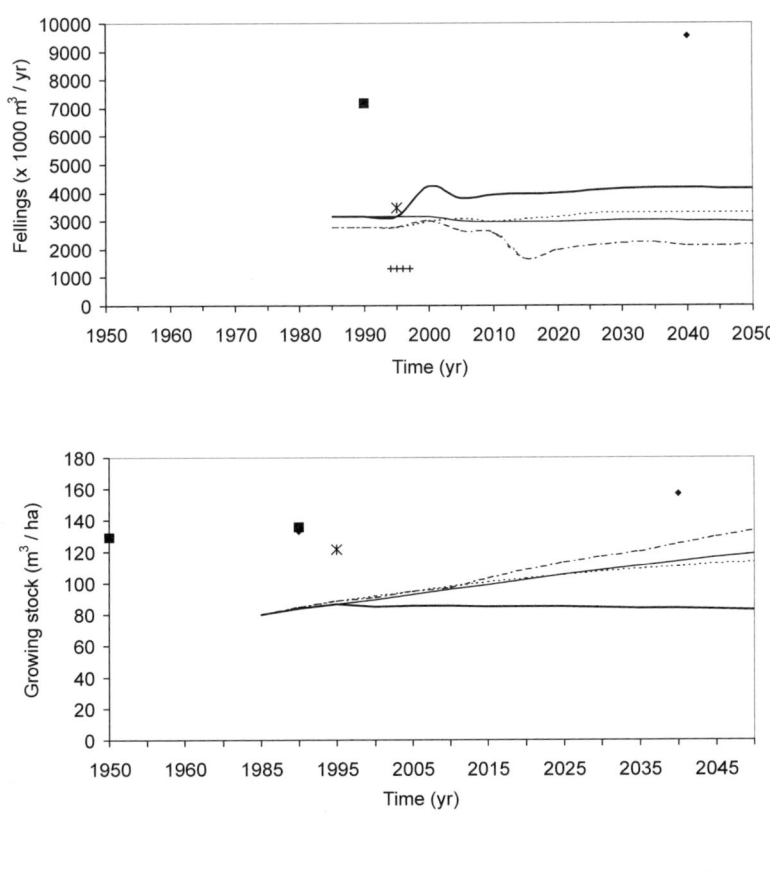

Fig. 5.86. Projected total national fellings (top) and growing stock per ha (bottom) in Yugoslavian forests under the four scenarios until 2050.

120 000 ha in coniferous high forest for the period 1985–2035. With the economic recovery underway (and a desire to attract tourism), it was assumed that there will be a strong emphasis on nature-oriented forest management. The rotation lengths were prolonged by 20 years in high forest and by 10 years in coppice. The share of thinnings from

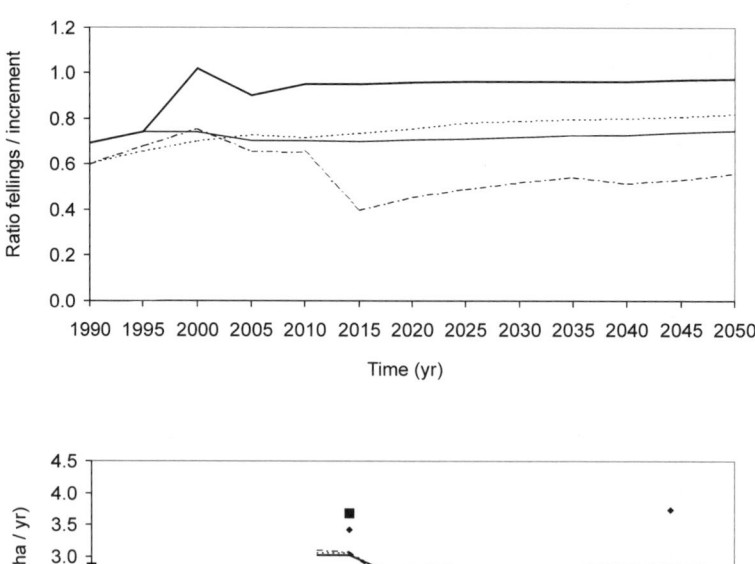

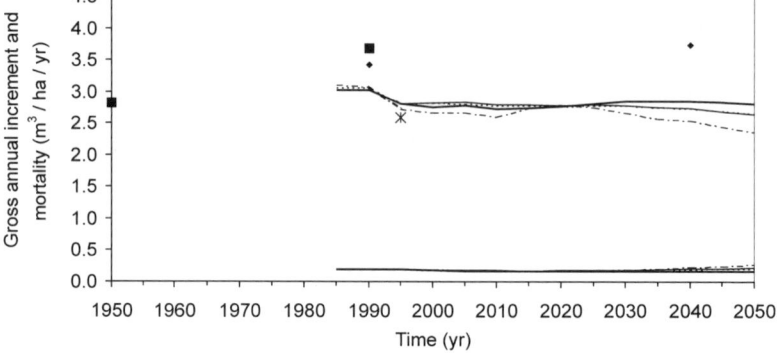

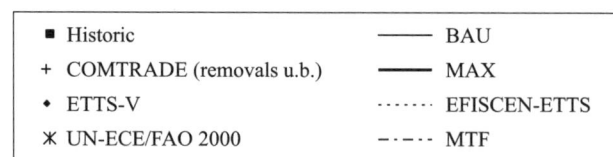

Fig. 5.87. Ratio of projected fellings/increment in Yugoslavian forests under the four scenarios until 2050 (top) and projected net annual increment and mortality (bottom).

total fellings gradually increased to 50%. We set aside broadleaved and coniferous high forests over 100 years of age and coppice forests of over 30 years of age. Initially this is an area of 594 000 ha. Through ageing this can increase during the simulation.

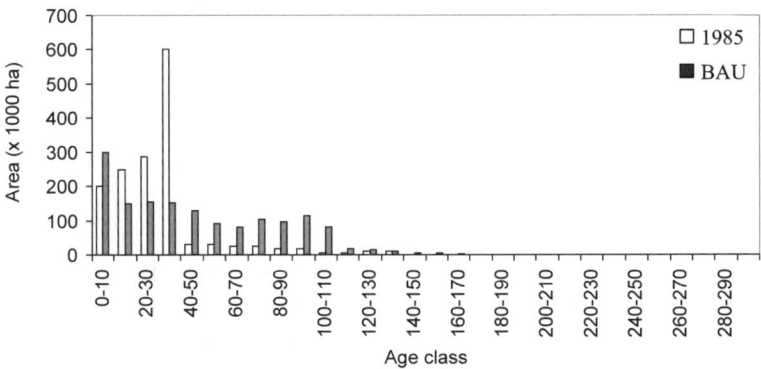

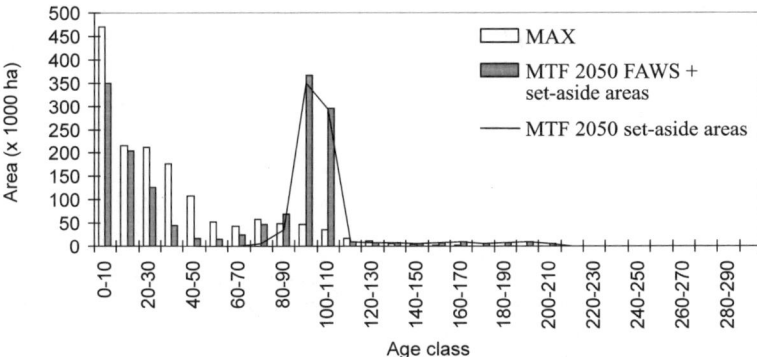

Fig. 5.88. The age class distribution of the Forest Available for Wood Supply (FAWS) in the initial situation and for 2050 under the BAU scenario (top). The age class distribution of FAWS for 2050 under the MAX scenario, the MTF scenario including set-aside areas and the MTF scenario with only set-aside areas (bottom).

Results for Yugoslavia

The results for Yugoslavia have to be regarded with care. This is because the underlying data are rather outdated, and because they only cover 65% of the forest area available for wood supply. Therefore, a comparison with international statistics is difficult. The most eye-catching results for Yugoslavian forests can be seen in the age class distributions for 2050 under the MTF scenario. A very prominent ageing can be seen in that scenario, as a result of the set-aside. The initial coppice area in the class 30–40 years, is completely set aside, and visible as reserves in the MTF age class graph for 2050. This also results in the fact that the required fellings cannot be found in the MTF scenario.

The projections for these forests in Yugoslavia show that a maximum sustainable production of 4.2 million m^3/yr may be achieved under the low increment as present in the data. Under a BAU scenario, much more evenly distributed age classes will develop and conversion of coppice to high forest occurs. The growing stocks increase to an average of 119 m^3/ha by 2050. Under this scenario the utilisation level of the increment then amounts to 75%.

In the MTF, mortality rates may then amount to 12% of the gross annual increment, producing 0.46 million m^3 of dead wood per year. The gross annual increment slightly decreases in all scenarios from 3.1 m^3/ha/yr initially, to around 2.7 m^3/ha/yr in 2050. The strongest decrease occurs in the MTF scenario, where the ageing is the strongest. In the MTF scenario 753 000 ha are projected to be newly established set-aside areas by 2050, leading to problems in finding the required fellings.

REFERENCES

Nilsson, S., Sallnäs, O. and Duinker, P. 1992. Future forest resources of Western and Eastern Europe. International Institute for Applied Systems Analysis. The Parthenon Publishing Group. England. 496 p.

UN-ECE/FAO 2000. Forest Resources of Europe, CIS, North America, Australia, Japan and New Zealand. Geneva, Timber and Forest Study Papers, No 17. 445 p.

6. EUROPEAN SCALE RESULTS

6.1. European totals

The projections mostly show a continuous build-up of growing stock, confirming the future continuation of the difference between increment and fellings. We foresee that total gross increment will slightly decrease to a level of around 4.8 m^3/ha/yr, providing a total increment varying between 637 million m^3/yr (MAX) and 729 million m^3/yr (MTF).

If fellings remain at the current total level of just under 400 million m^3/yr (BAU), then the average growing stock in Europe will rise to 226 m^3/ha in 2050 from the initial 137 m^3/ha in 1990 (Figure 6.1). This continuous rise levels off only slightly due to natural mortality of some 13% of gross annual increment in the Bus-as-us scenario in 2050.

Foreseen realistic felling increases (MTF) of some 0.3% per year do not curb this trend of increasing growing stock much. If the fellings increase to 450 million m^3/yr by 2030 as foreseen in the MTF scenario, then the average growing stock in Europe still rises to 218 m^3/ha in 2050. This is mainly because the felling/gross increment ratio in the MTF scenario is only 62% (or 70% of net increment). As can be seen in Figure 6.2 (bottom) this ratio has historically decreased in Europe since 1955. In 1955 it was 96%. UN-ECE/FAO (2000) reported a ratio of 53% for 1995 (well in agreement with our projection). The drop in this utilisation ratio is mainly because the increment has increased a lot in Europe since 1955.

The gross annual increment remains the highest in the MTF scenario (5.0 m^3/ha/yr in 2050) because in this scenario the largest proportion of total fellings comes from thinnings, leading to a larger proportion of the forest benefiting from some growth stimulation after thinning. This highest increment is remarkable because of the set-aside policy carried out under the MTF scenario. The ageing was slower than expected: from an average age of the European forest of 57 years in 1990 to only 68 years in the MTF scenario in 2050. This is only a few years older than what may happen under the BAU scenario. Once more confirming that it is difficult to influence the autonomous development present in a large forest resource is hard to change. The

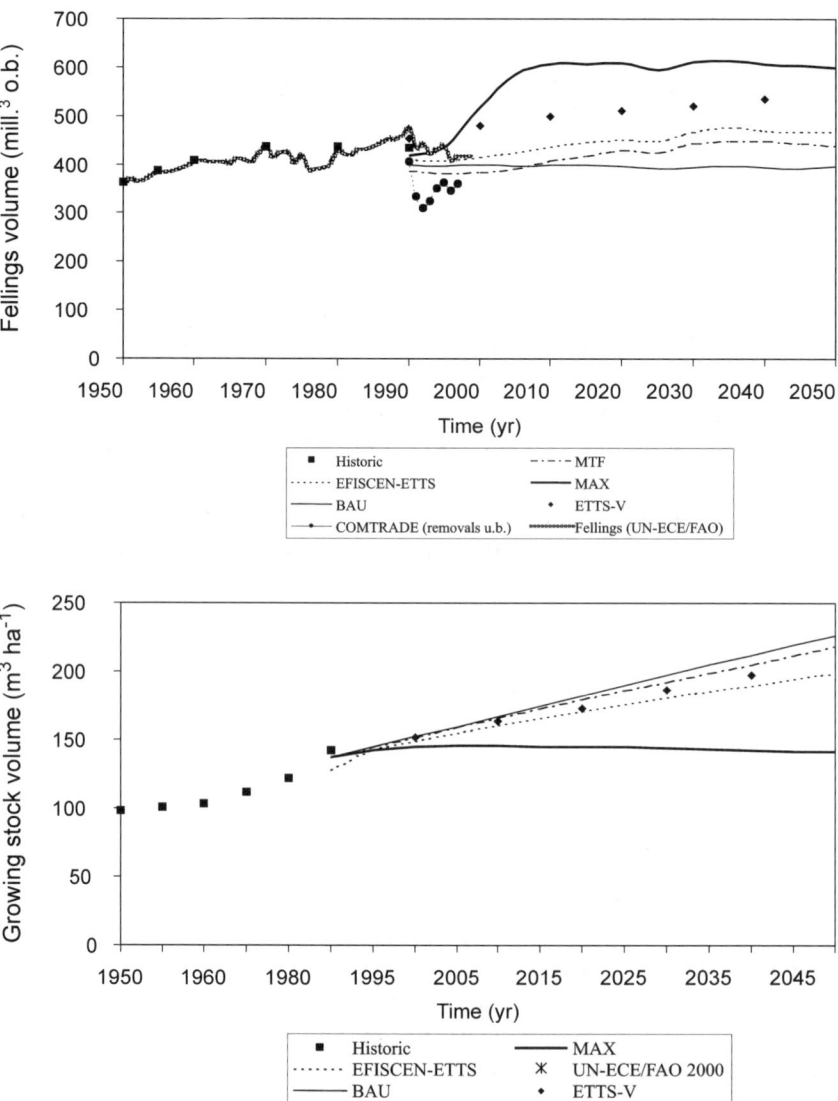

Fig. 6.1. Projected total fellings (top) and growing stock per ha (bottom) for the European forests under the four scenarios until 2050.

total area of reserves increased from 4 million ha in 1990 to 11.9 million ha in 2050. This did not hamper the identification of fellings

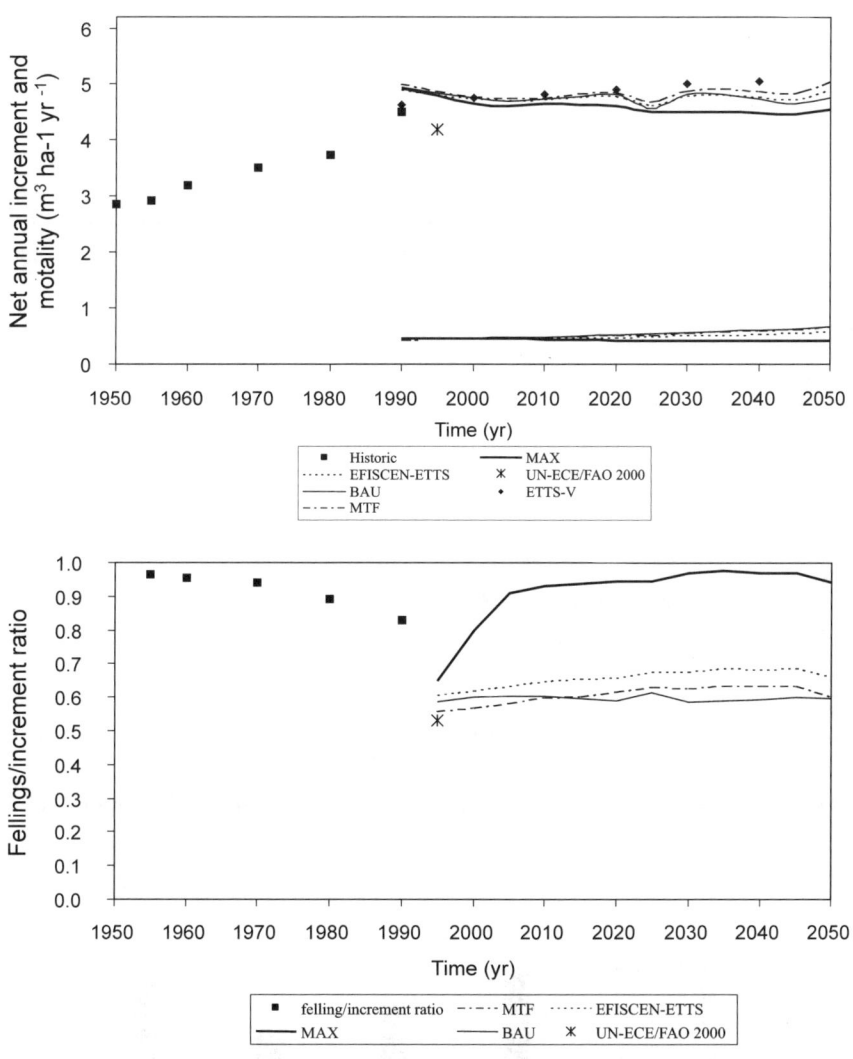

Fig. 6.2. Projected gross annual increment and mortality (top), and the felling/gross increment ratio (bottom) under the four scenarios. Historical, ETTS-V, and UN-ECE/FAO 2000 values are net annual increment values.

to fulfil the required fellings increase of 0.3% per year in the MTF scenario.

The projections show a maximum sustainable fellings level of 606 million m^3/yr (MAX in Figure 6.1 top; 95% of the gross increment, and 100% of the net increment) on the simulated forest

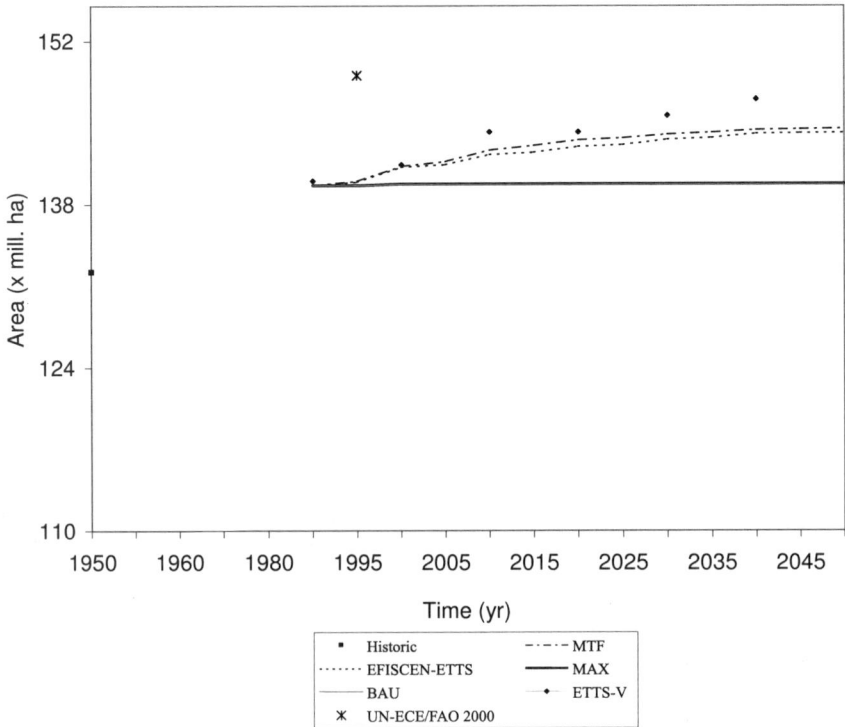

Fig. 6.3. European forest area and its expansion as simulated in the present study.

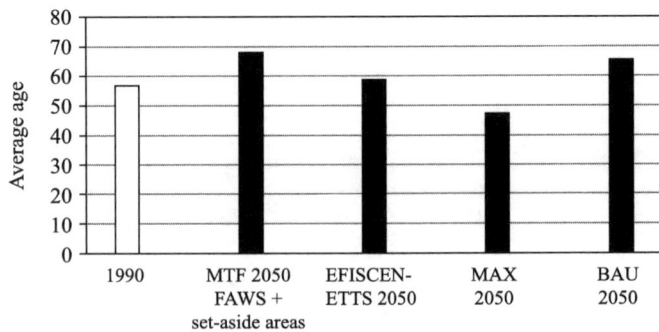

Fig. 6.4. Average age (years) of total European forests in 1990 (open bar) and in 2050 under the four scenarios (shaded bars).

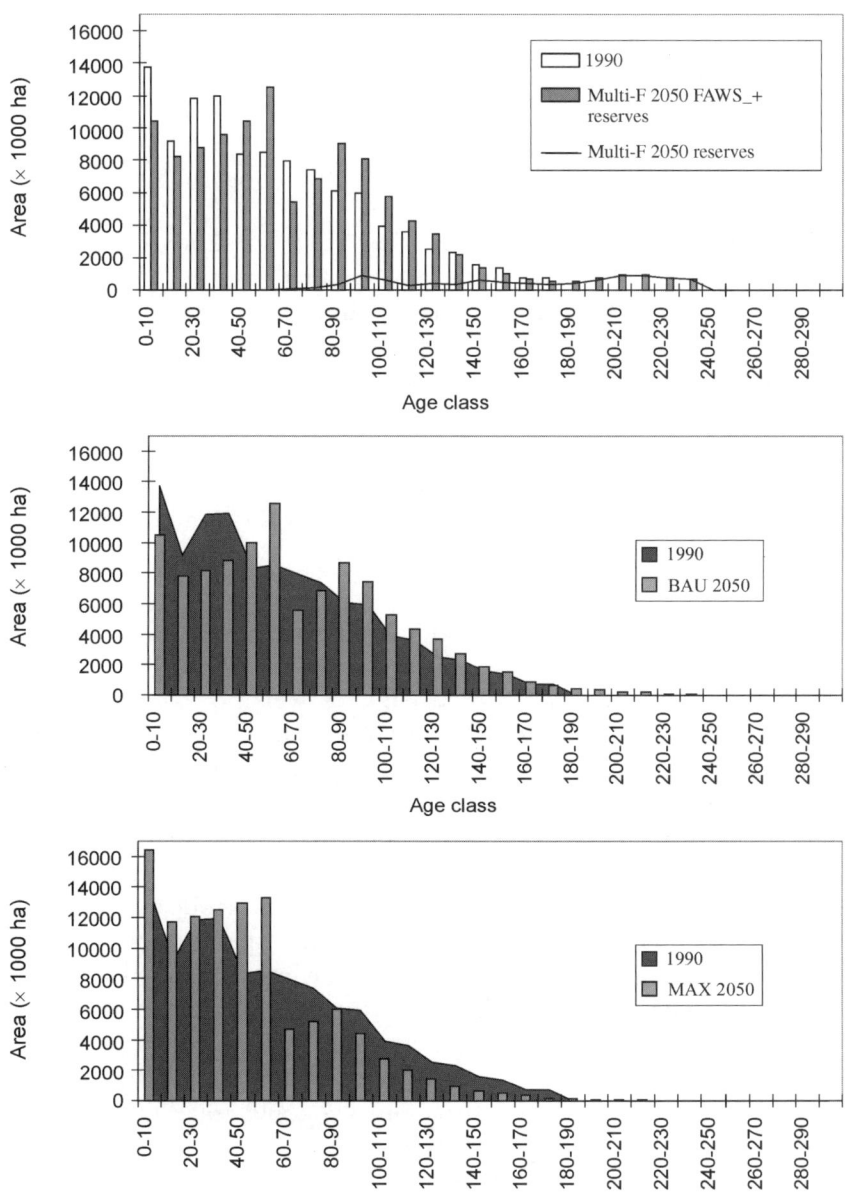

Fig. 6.5. Initial age class distributions and the situation in 2050 under three scenarios.

area of 139 million ha. Under that level the growing stock stabilises at around 143 m^3/ha.

The forest expansion carried out under the EFISCEN-ETTS and the MTF scenarios barely seem to have a noticeable impact on results. However, it should be noted that a sensitivity analysis as regarding maximum production levels with and without the forest expansion was not done here.

The strongest ageing occurs in the MTF scenario (Figures 6.4 and 6.5). This is partly due to the reserves that have been established in the course of 1990 to 2050. The continuous line in the top graph of Figure 6.5 shows that almost all the forest over 190 years has been set aside. This ageing is far less profound in the Bus-as-us scenario, although the growing stock volume increase was very noticeable in the BAU scenario.

The MAX scenario (especially) shows a profound shift in age classes towards younger classes. The average age under this scenario decreases from 57 years initially, to 47 years in 2050.

6.2. Problems in finding desired fellings

The aim of running the EFISCEN-ETTS scenario was to explore the comparability of output when running (in principle) the felling levels of another study. Therefore, we ran the felling levels of ETTS-V (Pajuoja 1995), and compared achievable felling levels from the present study with the ETTS-V output (Figure 6.6). The more dynamic approach used in the present study (ageing, management regimes with probabilities per forest type) presented significant problems (<0.85% of ETTS-V level found for 2040) in finding the fellings levels of ETTS-V for Belgium, Bosnia and Herzegovina, Croatia, Denmark, Greece, Ireland, Italy, Macedonia, Portugal, Spain and Yugoslavia. Furthermore, for the United Kingdom a significant problem in finding required fellings was found in the period 2010–2020. In the cases of Ireland, Portugal and United Kingdom we could trace the shortage back to large afforestation programmes, which take a long time before they reach harvestable ages. Therefore, caution should be used when making parallel investments in afforestation and in the processing industry. In the cases of Bosnia and Herzegovina, Croatia, Greece, Macedonia and Yugoslavia the shortage can be traced back to overall uncertainty in underlying data. For the remaining countries (Belgium,

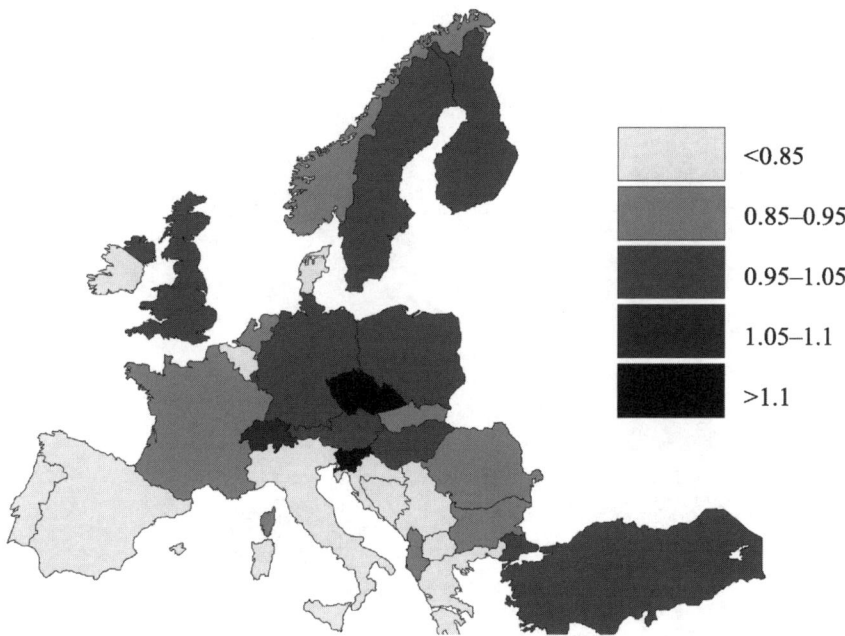

Fig. 6.6. Fraction of ETTS-V fellings achieved in EFISCEN-ETTS scenario for 2040.

Denmark, Italy and Spain) the reported shortage in finding required fellings is a combination of the way management regimes were set and the development of age class distributions. For this group it is possible that owners would sufficiently (and dynamically) adjust their harvesting behaviour as soon as a shortage starts to occur (higher prices for roundwood). This dynamic (price elastic) behaviour of owners is, however, not part of EFISCEN yet. Nevertheless, our results indicate possible future problems in finding the required felling levels.

6.3. Spatial distribution of the results

The maps in Figures 6.7 and 6.8 show that the build-up of growing stock occurs in almost all countries of Europe in the MTF scenario. The build up is concentrated mainly in the Atlantic and Central European countries. The highest gross annual increments (GAIs) are also found there. But these GAIs tend to decrease slightly. The combination of high GAIs and moderate to average utilisation ratios

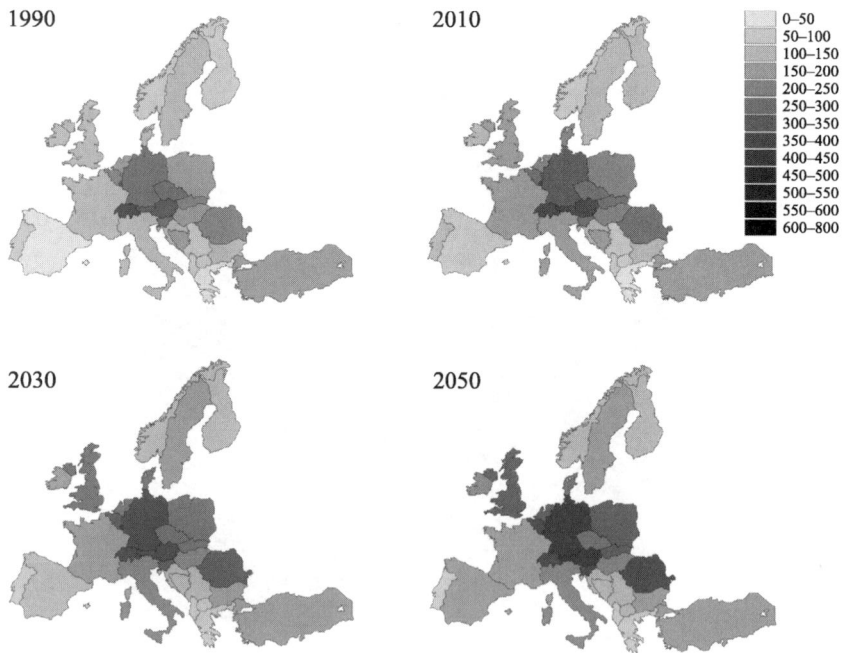

Fig. 6.7. Growing stock (m^3/ha) development for each country in the MTF scenario through time.

in Atlantic and Central Europe, results in the spatial distribution of the growing stock as shown in Figure 6.7.

Figure 6.9 shows that the MTF and the BAU scenarios differ very little in the spatial distribution of the growing stock build-up, and in the absolute values of the growing stock. The MAX scenario keeps the national average growing stocks at approximately the same level as in the top left map of Figure 6.7.

As we had seen from Figure 6.4, the average age of the total of European forests did not change dramatically under any of the scenarios (e.g. the largest change was from an average age of 57 years in 1990 to 68 years in 2050 in the MTF scenario). This is reflected in Figure 6.10, which shows the ageing for each country between 1990 and 2050. The strongest ageing is again visible in the BAU scenario. There are no large opposite trends between countries that might have resulted in the small European average ageing. Only in the MTF scenario did the forests in Ireland and Portugal hardly age. This was due to large afforestation programmes.

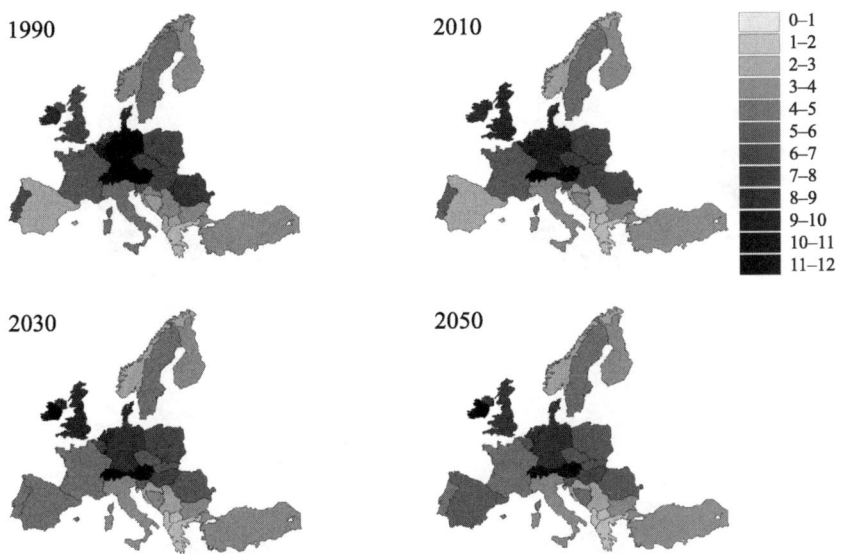

Fig. 6.8. Gross annual increment (m³/ha/yr) development for each country in the MTF scenario through time.

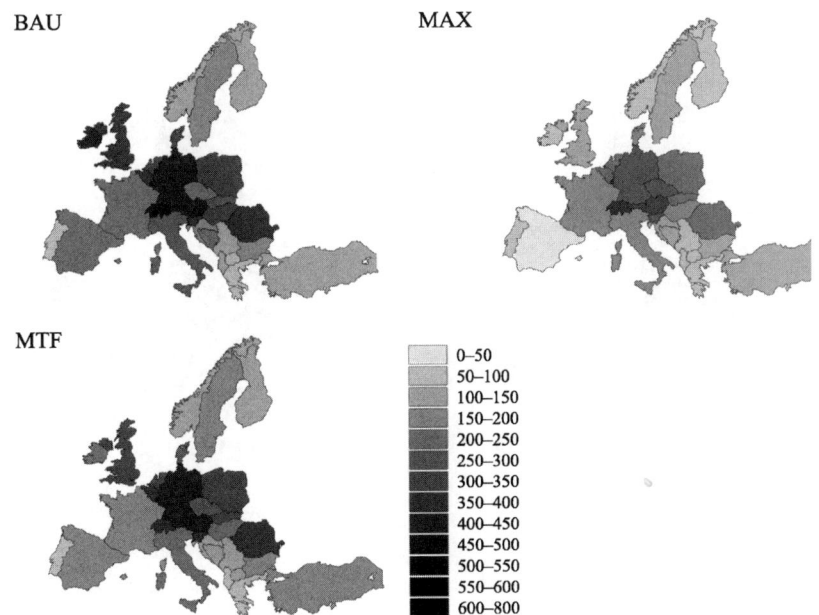

Fig. 6.9. Average growing stock (m³/ha) for each country and scenario in 2050.

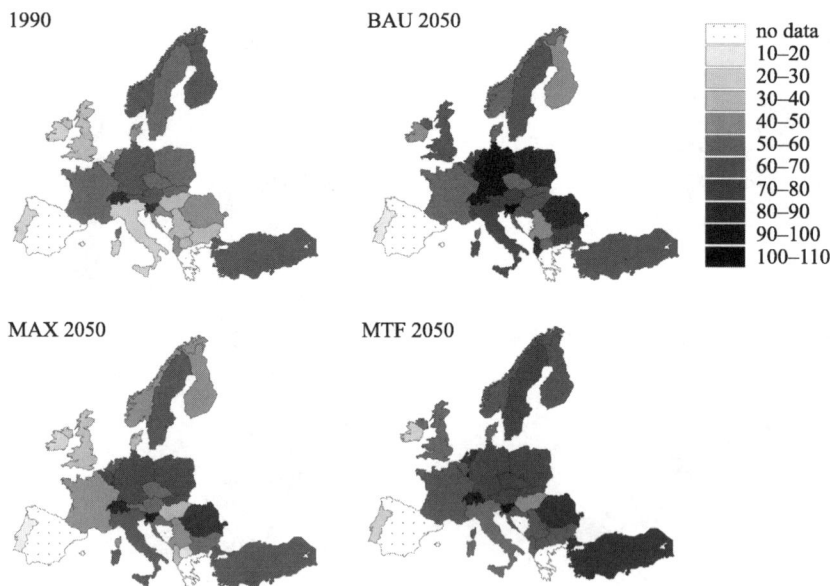

Fig. 6.10. Average age of the forests in 1990 and in 2050 for each country and scenario (Spain, Greece and Bosnia and Herzegovina are not given because either a simple approach or the diameter distribution approach was used).

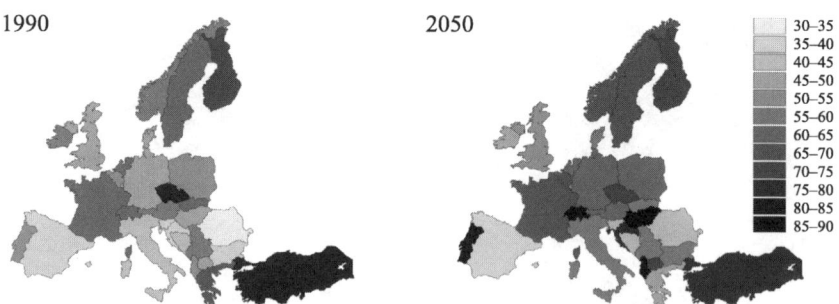

Fig. 6.11. Felling/gross increment ratio in 1990 and in 2050 in the MTF scenario.

Figure 6.2 (bottom) showed that the felling/gross increment ratio slightly increased in the MTF scenario for Europe as a whole. This is visualised for each country in Figure 6.11. The strongest increase in time in felling/gross increment ratio (more than 20%) was found for Albania, Croatia, Hungary, Macedonia, Portugal and Switzerland.

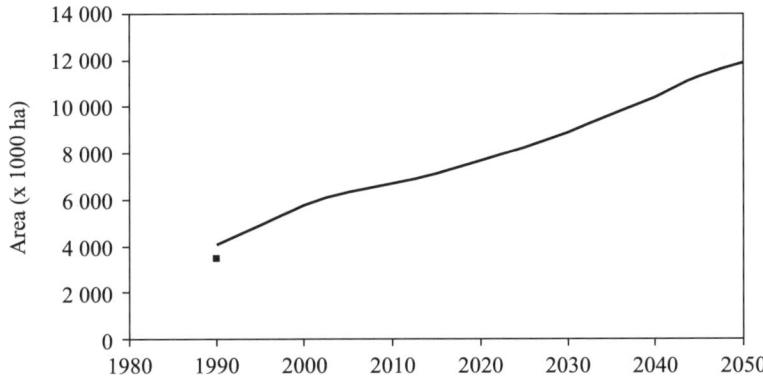

Fig. 6.12. Development of the set-aside forest areas under the MTF scenario. The square given for 1990 is based on statistics from the countries as presented by Diaci (1999).

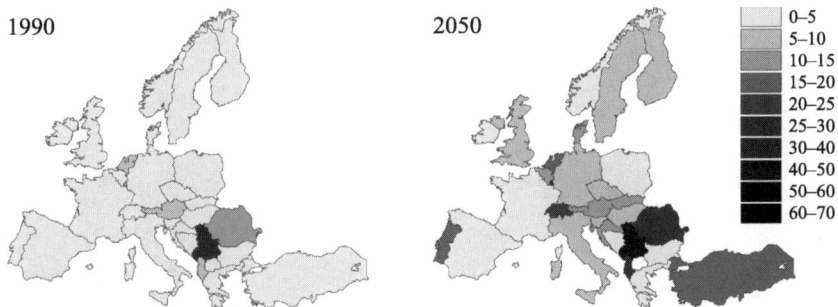

Fig. 6.13. Percentage of forest area that is set aside from production in the MTF scenario in 1990 and in 2050.

In absolute terms high felling/gross increment ratios (more than 85%) were found for Albania, Hungary, Portugal and Switzerland.

The establishment of set-aside forests by assigning a probability of zero to final harvest of certain age classes, leads to a gradual increase in this set-aside area because of ageing during the simulation (Figure 6.12). By 2050, the total area of these types of reserves has reached 11.9 million ha. Partly due to this set-aside and partly due to ageing and increase in growing stock, the dead wood production increases significantly in the MTF scenario to a total production of 83.4 million m^3/yr in 2050.

Because of the unpredictability of the ageing process the establishment of set-aside areas is concentrated in the eastern Mediterranean region. A concentration in the highly urbanised societies of the Netherlands, France, Germany, Belgium, Luxembourg and the UK would seem more realistic.

7. DISCUSSION

7.1. Uncertainties

The results presented in the present study are projections of 'what if' scenarios. Therefore, they should not be seen as predictions of the future, but as an indication of what may happen under alternative regimes or policies. Large-scale and long-term projections for a large number of countries, as has been done here, certainly carry disadvantages. Specific local or national circumstances can never be taken into account as accurately as would have been done when a separate study had been carried out for that locality only. Therefore, the results as presented here are not meant to replace national level studies, but are a multi-national strategic level study that calculates and presents results based on a harmonised methodology. This makes comparison of results between countries simpler.

Simulations of future development of forests are surrounded by uncertainty. This overall uncertainty includes uncertainty in the input data, inaccuracy of the growth predictions, model assumptions, and the assumptions underlying the scenarios. Therefore, the certainty with which the results can be regarded will vary from country to country, and the uncertainty will increase the further we progress in the simulation. In this chapter, the different aspects of these types of uncertainty will be discussed.

7.1.1. Initial state

Practically all European countries have a regular forest inventory that is updated in cycles (usually every 10–15 years). Exceptions to this are Albania, Bosnia and Herzegovina, Greece, Macedonia and Yugoslavia. The input data that could be obtained for these countries are believed to be a rather poor reflection of the current state of the forest. The data are either outdated or reflect only a small part of the total forest area in the country. Chapter 3 and Figure 7.2 present a full account of the share of forest in each country that was covered in the present study. The five countries mentioned above will not have a large influence on overall uncertainty, because these countries account for only 6% of the total European forest area.

The aim of inventories was traditionally to inform government, forest owners, industry and the public about the state and potentials of the forests. In Western European countries this was done through a sampling based inventory, and in Central and Eastern European countries through a complete stand survey. Sampling errors can therefore be delivered only for sampling inventory for Western European countries. Stand-based inventories include error due to assessment method and are exposed to bias.

For the first group, the reported sampling standard errors (s.e.) are small, up to 1.5% at the national level for forest area, and up to 5.1% at the national level for growing stock. The few countries that report standard errors at the national level for increment (Finland, France, Norway, Sweden and Switzerland), report standard errors of 0.4–1.4% (Laitat et al. 2000). However, sampling designs are set up to provide accurate information at the national level, so when countries report results at the provincial level, then standard errors rise quickly to up to 10%. When results are presented at a further level of detail (per species and age class) standard errors rise even more. This means that in the present study where we work with detailed results of national inventories, we start off with rather high uncertainties at the single age class level. However, at the national level (the level we report at in Chapter 5), the standard errors will be small again.

Uncertainty is also caused by differences in the initial year of the input data. On average the inventory data that were used for the initial year present the state of the European forests of 1988 (Figure 7.1). However, this varies from 1974 to 1995. When an inventory year that was not rounded to '5' was reported (e.g. 1987), we simply rounded it to the nearest '5' without adjusting the data (i.e. 1985). Runs were made from the initial year to 2050, but results were only presented from 1990 onwards (except for Hungary and the Slovak Republic that had 1995 as the initial year).

Sample-based inventories carried out in Europe vary with regard to the definitions and methods used. All together this may yield adjustments required when the same definitions and methods would be used in all countries (Köhl and Päivinen 1997, UN-ECE/FAO 2000). We disregarded this aspect in the current study. We simply used the area, growing stock and (the assumed gross) increment data

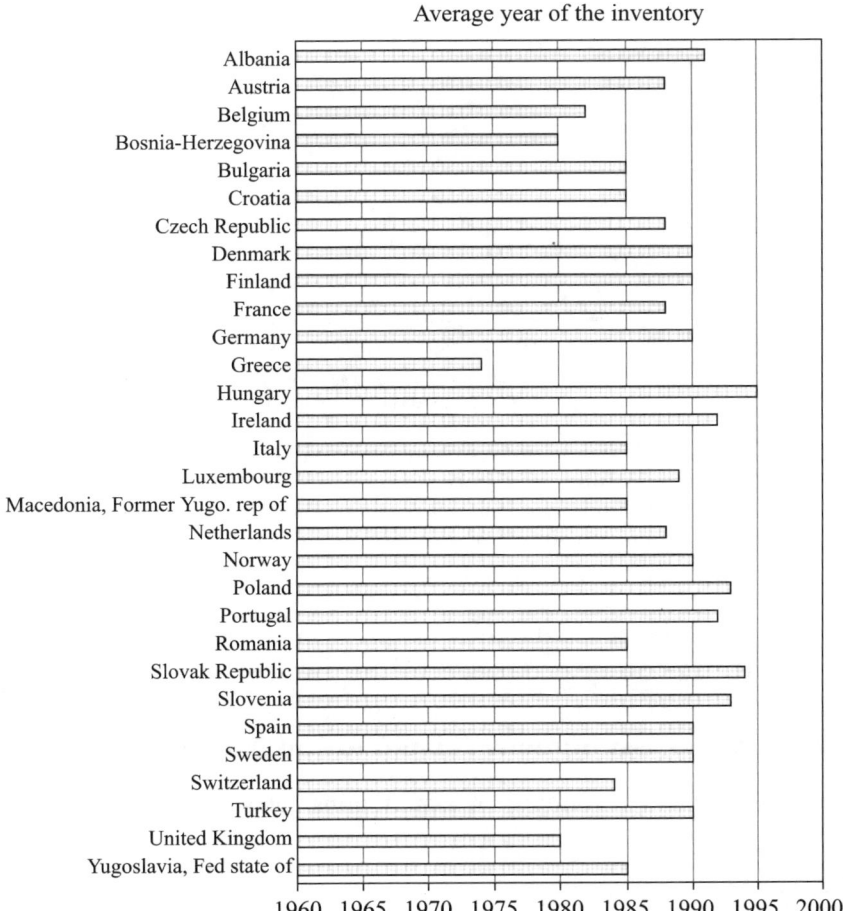

Fig. 7.1. Initial year of the simulations for each country as determined by the average base year of the specific cycle of each national inventory as used in the present study.

as reported by the country to us in 1996 for this study. The only exception is the Czech Republic that reported underbark data. These data were converted to overbark data by multiplying by a factor 1.14 for forests older than or equal to 55 years, and by a factor 1.18 for forests younger than 55 years.

Usually the data as supplied by the country correspondents represented the exploitable forest area according to UN-ECE/FAO (1992)

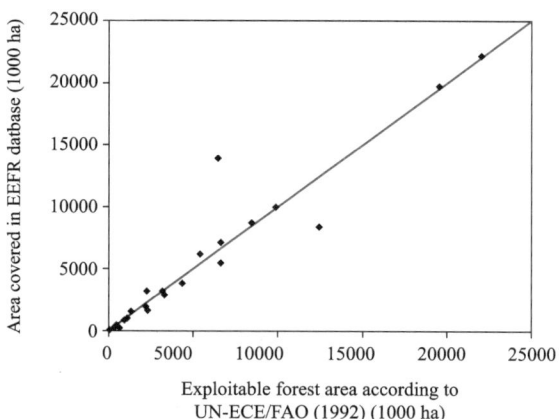

Fig. 7.2. Area covered in the present study vs. the forest area available for wood supply according to UN-ECE/FAO (1992). The outlier above the y = x line is Spain, and note that for some of the small forested countries, the relative deviation can sometimes be large.

(Figure 7.2). Sometimes the area of special forest types were lacking because the age class based data could not be delivered.

7.1.2. Basic assumptions related to simulations

The study makes a number of assumptions related to the following topics:

Resources

It is assumed that:

- the forest inventory data supplied by countries represent the state of the national forest in sufficient detail;
- a major forest area expansion will not occur; and
- generation will be carried out using the same species as the one that was clearcut, and thus a species distribution change will not take place.

Growth

It is assumed that:

- forest growth can be simulated sufficiently accurate at the minimum area of 10 ha for each cell in each matrix; and that

- environmental changes such as climate change, pollution, CO_2 fertilisation and N deposition will not change the site fertility during the simulation period. Thus, growth rates will stay at the level of the 1980s and 1990s on which the growth functions were fitted.

Mortality

In relation to environmental changes, we assume that large-scale decline does not occur. Natural mortality will remain a function of density, age and growing stock volume. There will be no change in disturbance rate, and practically all occurring natural disturbances will be salvaged and taken up in planned harvesting as has been the case in the past.

7.1.3. Simulated results for increment and mortality

Many of the national level results on gross increment show a slightly declining increment over time (of an average from 4.95 m^3/ha/yr in the initial year to some 4.8 in 2050). This seems strange especially in respect to the historic trend of increasing increment. The EFISCEN model suggests that the permanence of the historic increase in increment would slow down as a function of ageing of the forest and the high growing stocks being reached. No doubt, there is uncertainty in the increment decline as given in Figure 6.2 (top).

The increase in increment in the historic data is partly caused by improved inventories over time, but also partly by stands reaching higher growing stock, and by enhanced growth rates due to improved management, and possibly CO_2 and nitrogen fertilisation (Spiecker et al. 1996). At some point in time, additional growing stock will not lead to increased growth rates anymore.

Since the increment is somewhat uncertain, this is reflected in uncertain growing stock development in the long term. Thus the increase in growing stock given in Figure 6.1 (bottom), may level off sooner (in case mortality rates would be higher and/or increment would be lower). One of the main research issues in developing the model further is when, and at which level the growing stock will level off in different forest types and management regimes.

The mortality was incorporated in a standard way for all countries because of a lack of reliable data. Mortality was parameterised as a 1% chance for a forest area to move one volume class down per

5 year simulation step for all forests up to 150 years of age. Above this age, the chance was increased by 0.25% for every 10-year age class. In addition this chance was increased by a factor 3 in case a forest area reaches the top volume class. Usually this resulted in mortality rates of 7% of the gross increment in the initial year. This is in accordance with UN-ECE/FAO (2000) which reports mortality rates of 8% of the gross increment for the same 30 countries. The way mortality was simulated (as taken from the increment) also reflects our assumption that the increment data as received from the country correspondents were considered gross increment data. This assumption was mainly based on a comparison of the current increment data to UN-ECE/FAO (2000) increment data. The latter are all displayed in the increment graphs as a single star in each country section in Chapter 5.

There are few other projections and international statistics for European forests. Our simulated increments are slightly higher than UN-ECE/FAO's (2000): The gross and net annual increment for the first period for the area of countries in our present study were 4.95 respectively 4.5 m^3/ha/yr, and those of UN-ECE/FAO (2000) 4.6 respectively 4.2 m^3/ha/yr. Most of the other projections used/present a lower increment (Table 7.1). Nilsson et al. (1992) based their

Table 7.1. Comparison of projections and a statistic made for European forests.

	Area included (million ha)	Growing stock (m^3/ha)	Annual increment (m^3/ha/yr)	Fellings (million m^3/yr)
IIASA Forest study (baseline) (Nilsson et al. 1992)	139	2080: 166	4.4 (average for 100 year simulation)	2060: 548.4
IIASA forest study (decline scenario) (Nilsson et al. 1992)	139	2080: 164	3.7 (average for 100 year simulation)	2060: 475.0
ETTS-V (Pajuoja 1995)	140	2040: 197	2050 NAI: 5.05	2050: 534.3
Statistics: UN-ECE/FAO (2000)	149 (FAWS)	1996: 143	1996 NAI: 4.2	
Present study (multi-functional scenario)	139	2050: 218	2050 GAI: 4.8 2050 NAI: 4.2	2050: 441

simulations on forest inventory data of the late 1970s and early 1980s. That was a time when not all European countries had a proper forest inventory. Often, before a proper inventory was available in a country, the increment estimates were very modest. As soon as the first inventory in a country became available, it usually showed a much higher increment than assumed (Kauppi et al. 1992). Furthermore, Nilsson et al. (1992) gave one estimate for their whole simulation period, interpreted here as a straight line through time. The study by Pajuoja (1995) presents a comparable (net) increment for 2050 to our results. However, he had a lower net annual increment for the initial year (4.6 m^3/ha/yr). That is partly due to the same reasons as in Nilsson et al. (1992), i.e. rather old increment estimates. But Pajuoja (1995) often presented an increasing trend for the projected increment. This can only be explained by the method he used, namely questionnaires sent out to country correspondents. The correspondents often used very simple projection methods, not taking into account the level of growing stock, or ageing of the forest, and possibly introduced model bias.

7.1.4. Model behaviour and validation

On the aspect of model behaviour, it is clear that the two other modelling approaches (Figure 7.3) used in the present study (diameter distribution approach and the simple balance approach) yield a different trend in results (see e.g. increment in Spain). This again underlines the importance of the core modelling assumptions, and thus introduction of possible model bias. With regard to this aspect, improvements will be possible especially in the diameter distribution approach.

The accuracy of the even-aged part of the projections made by EFISCEN's even-aged area matrix approach has been described in Sallnäs (1990) for the growth models and in Nabuurs et al. (2000) for projections for a country as a whole. In the latter study, the EFISCEN model was parameterised on inventory data of Finnish forests for the 1920s and the model was run until 1990. Figure 7.4 presents the main result of that study with the unimproved model version that used net annual increment as modelling approach. The validation part of the study (for projections over the period 1923–1990) showed a serious underestimation of future increment when growth functions were parameterised based on the 1921–1923 inventory data. This

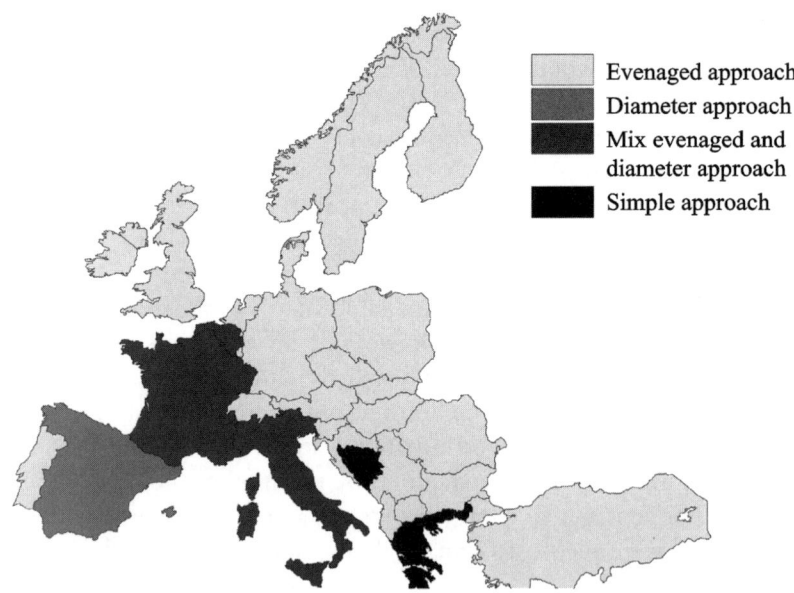

Fig. 7.3. Approach used for each country – light gray: even-aged approach, gray: diameter distribution approach, dark gray: mix of even-aged and diameter distribution approach, black: simple balance method.

underestimation was mainly caused by the fact that as of 1960 the net annual increment in Finnish forests started to increase as a result of site improvement and higher stocking. The underestimation led to a combination of deviations occurring in growing stock development and in age class development. This resulted in the shown deviation. Later, in Nabuurs et al. (2000) EFISCEN was improved concerning increasing net increment after thinning, ageing of stands, growth rates at high growing stocks, and natural mortality. Incorporation of transient changes in growth rates was only applied in Kramer and Mohren (2001), and Nabuurs et al. (2002).

It was then concluded that EFISCEN can accurately project the forest development for a period of up to 50 years. After that deviations may still accumulate and affect each other through feedback mechanisms.

A comparison of EFISCEN national level output with other projections was made in Nabuurs et al. (1998). They concluded that EFISCEN is able to reproduce the ETTS scenarios in a satisfactory way. In that study, EFISCEN results were compared with ETTS-V

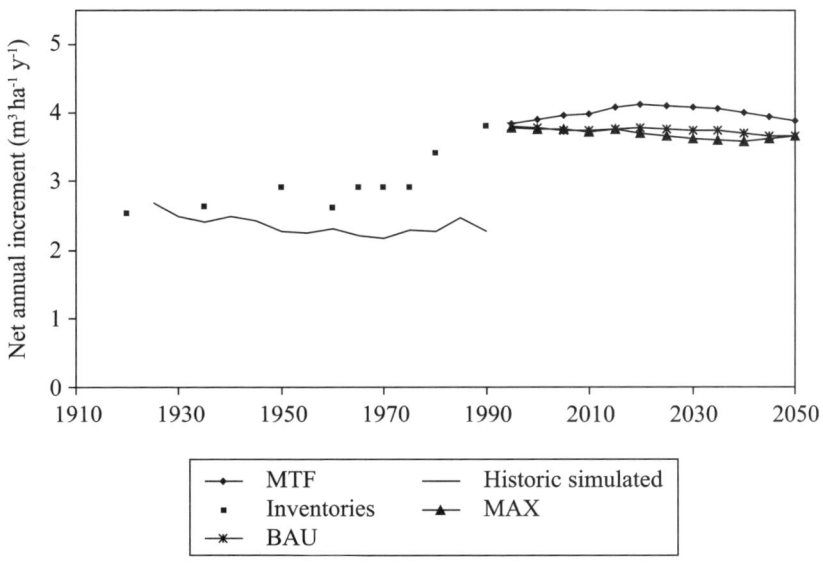

Fig. 7.4. Development of mean net annual increment of Finnish forests of both the validation part of the study (1923–1990) and the projection part (1995–2050).

results (Pajuoja 1995). Where differences in output occurred, they were explained either from differences in input data, growth models, or by the fact that a more dynamic approach was incorporated in EFISCEN. This more dynamic approach takes into account the development of growing stock volumes and age of the forest and thus increment of the forest in a better way. Also, EFISCEN provides output on more variables and at several levels of detail.

Comparison by Nuutinen and Kellomäki (2001) showed significant differences when projecting Finnish forest resources with the MELA, SIMA and EFISCEN models. However, for their comparison they used the Nabuurs et al. (1998) study only, not the improved version of Nabuurs (2001), Pussinen et al. (2001), or Kramer and Mohren (2001). The comparison showed that model bias (in whatever form or in whatever model) can be a serious problem. Built-in assumptions on relative growth rates compared with growing stock determined the outcomes of the various models to some extent.

In Belgium, Bosnia and Herzegovina, Croatia, Denmark, Greece, Ireland, Italy, Macedonia, Portugal, Spain and Yugoslavia it was not possible to meet the demand assumptions of ETTS-V (<85%

of ETTS-V felling level found in present study for 2040), given the growth rates and the management regimes applied in this study. This resulted in a reduction of European fellings level in the present study of 62.6 million m^3/yr compared with the actual ETTS projection. This, together with a lower GAI (and thus lower NAI as well) in the present study resulted here in a lower average growing stock of 7 m^3/ha in 2040. In some countries the gross annual increment development as simulated by EFISCEN was very different from the results of ETTS-V. This was the case for Finland, Macedonia, Norway, Poland, Slovenia, Switzerland and Yugoslavia.

8. EUROPE'S FOREST RESOURCE POTENTIAL AND POLICY IMPLICATIONS

In co-operation with Ilpo Tikkanen

8.1. Summarising the results

This study has concentrated on describing alternative futures for the forest resources in Europe. However, the future of forests might differ greatly from the scenarios presented here due to, for example, changes in socio-economic factors, policies influencing management regimes and climate. In the following chapter, a few issues related to policy options and these underlying driving forces are discussed.

One main point for understanding policy options is to realise that the current state of the European forests is the result of management over the last couple of centuries. That has created a resource with certain characteristics (species, age class distributions, etc.) that can be subject to changes over time. However, a forest resource is an inert system with 'memory' of the past management.

Another point for understanding policy options is that there is no such thing as 'common European forestry'. Its main characteristic is the wide diversity at the national level of laws, state of the forest resource, aims, and culture of management.

- The business-as-usual scenario showed that the present felling level can be maintained in the future in a sustainable way. However, it will lead to denser and probably unstable forests, or, when regarded from another point of view to potentially higher biodiversity values in the long-term through the occurrence of more natural disturbances and mortality. Under the simulated growing stock for 2050 under the BAU scenario, the impact of disturbances and natural mortality may increase, reducing the net annual increment, and therefore endangering the long-term sustainability of this scenario. Therefore, it is not certain that the average growing stock of around 240 m^3/ha in 2050, which is presented for three rather modest fellings scenarios, will actually be achieved.

- The maximum sustainable production scenario demonstrates a theoretical ceiling of the potential of existing forest resources under current management regime and restrictions. Fellings go up in the countries covered by this study by 195 million m^3/yr at which the felling/increment ratio is 95%, the remainder being a small build-up of growing stock and natural mortality.
- The multi-functional scenario tried to combine the wish to produce more wood along with a higher value placed on nature. The area of set-aside forests was increased. The main finding was that with 11.9 million ha of set-aside area and a total production of dead wood of 83.4 million m^3/yr in 2050, the desired fellings increase could still be achieved. However, this scenario still leads to denser and maybe unstable forests, which may lead to undermining of the increment in the long term.

8.2. Forest resource development in a socio-economic and policy context

A status quo of forest resource is an outcome of: (1) market forces (demand); (2) cutting and investment behaviour of forest owners (supply); (3) ecological processes of forests themselves; and (4) the impacts of implemented forest policies and other policies influencing forests and their use and development (see Figure 8.1).

The present state of the forests leads, as a result of these inter-linked elements and biological processes, to the prospective forests in

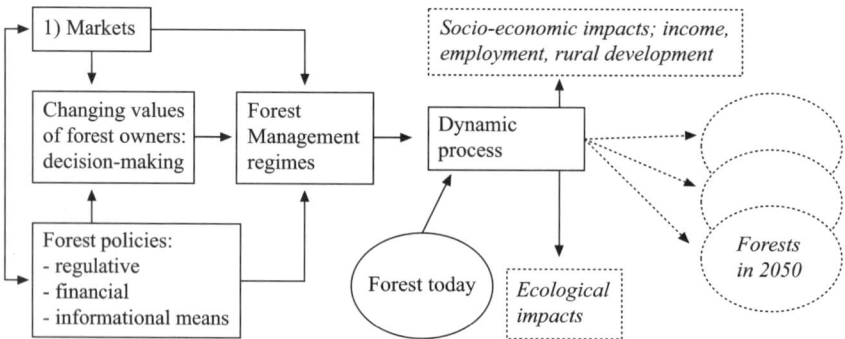

Fig. 8.1. Role of markets, forest owners' behaviour and policy factors in the development of the forest resources.

2050. The business-as-usual scenario provides a general starting point for policy deliberations about available policy options. It is based on the assumption that there will not be any significant changes in these underlying factors and their trends, and policies remain unchanged. The business-as-usual scenario will continue the past trends of forest utilisation and will lead to a constant accumulation of forest resources. At the European scale, only slightly more than half of the gross annual increment will be harvested on average. When compared with the maximum sustainable production scenario, the utilisation rate continues to remain rather low.

At first glance the scenarios do not provide any dramatically new outlooks for the future forests in Europe, as compared with other studies. However, from the policy formulation perspective the information obtained from the scenarios contains a very important message and the main conclusion:

> *European forests and their potential provide room for policy-making to design socially, economically and environmentally balanced policy options.*

Within the limits of sustainable forest management there is space in most of the countries for increased economic utilisation of cutting possibilities and at the same time ensuring the nature conservation objectives. It must be emphasised, however, that the situation varies from country to country.

The increasing resource provides policy makers with a challenge (but also the luxurious situation) to choose between varying combinations of aims for biodiversity values and increased cuttings for rural development and employment opportunities according to national priorities.

8.3. Shaping the policy options for the future

From the market side, it seems that in the foreseeable future the demand for wood, as well as non-wood goods and services, is increasing at the European scale (see Trømborg et al. 2000). The main question is how (non-)market forces, forest management decisions of forest owners and prevailing forest policies respond to these demands.

From the policy side, the development of global and European forest policies during the 1990s form the frame and starting point for future policy options. The outcome of these policy formulation processes and discussions can be seen in several policy-making aspects:

- Forest policy that aims towards sustainable forest management has become widely accepted, and the development of criteria and indicators provide evolving tools to further specify policy goals as well as to monitor the progress towards these goals;
- New global policy instruments, both legally binding and non-legally binding, such as the Convention on Biodiversity and The United Nations Framework Convention on Climate Change, have been established;
- International and Pan-European processes and institutions, especially IPF/IFF-processes and Ministerial Conferences on the Protection of Forests in Europe (MCPFE) have produced several politically important resolutions, recommendations and proposals for action. A new institution, UNFF, with a future oriented programme of work, has recently been established;
- New strategies for the forestry sector have also been launched, such as the Forest Strategy of the European Union.

The outcomes of these international policy processes have also reflected on national level policy-making in terms of renewed national forest programmes, legislation, and development of criteria and indicators in several countries in Europe.

Given this frame as a starting point to shape the future policy options for forest resource utilisation and development there are a few major challenges that policy-makers are facing in the short term:

1. How to co-ordinate various policy instruments, processes and initiatives; and
2. How to implement the existing mix of policy means and programmes.

In designing future forest policies to take advantage of the room for policy options that the potential of forests provide to the benefit of European people and societies we may raise the third challenge to be discussed between policy makers and research networks:

3. How best to integrate relevant research and knowledge as part of policy development processes.

Thus, the question of how to design the combination of many goals with varying preferences among many stakeholders, in order to achieve sustainability locally, nationally and internationally, asks for wise policy making processes based on sound research.

8.4. What should we know to control the development?

Looking at the options that the potential of existing forest resources provide, we are facing other types of challenges: knowledge of policy alternatives and their impacts.

The underlying assumptions of the Business as usual scenario cover various driving forces influencing forests of which the existing socio-economic and policy knowledge in support of policy-making is still limited. What type of knowledge would be relevant and needed, in addition to the information produced in this study, to shape the future policy options?

Three major categories of knowledge requirements and Pan-European research needs can be identified:

- Changing values and forest management behaviour of forest owners.

Changing values and preferences of forest owners affect forest management decisions, cutting and investment behaviour, and therefore, determine the long-term supply responses to increased demand. These behavioural studies are important in two aspects: How do forest owners react in their forest management decisions on: (1) markets; and (2) forest policy instruments and programmes? This information would provide an indication of the potential shifts in forest resource utilisation and development in the long term and reveal potential policy implications about whether the likely trends are desirable from the welfare aims point-of-view or not.

- Socio-economic impacts of different forest management regimes.

The maximum sustainable production scenario as one policy option shows a rising trend of cutting potentials for rural income, employment and economic development, enhancing the raw material base for forest industry investments as well as increasing the potential for bio-energy use of forests. Policy shifts towards this option would also enhance the national economic impacts of the forestry sector, but will have to be achieved along with environmental and social pressures to continue set-aside policies. Social values favouring the policy shift further in this direction may set, in extreme cases, a constraint on the expansion of the forestry sector in the future. Policy deliberations between these two options would require further analysis of socio-economic and ecological impacts of different forest management regimes as an outcome of practical implementation of alternative policy programmes. These analyses should cover impacts both at enterprise and national economy levels.

- Impacts of various forest policy instruments and their mix on forest management practices.

The projected scenarios of European forests provide a fundamental question for policy-making in European societies: are business-as-usual policies preferred by policy-makers in the future, or, is there a need to redesign policy interventions to contribute to the utilisation of these potentials for the welfare of European people and sustainable development? Policy discussions during the 1990s show that in changing socio-economic and environmental conditions the status-quo policy option may not be the preferred one. In circumstances where new and innovative policy formulations are called for, a second series of questions about 'What should we know?' will arise: What are the impacts of alternative policy means? How effective and efficient are regulative, financial or informational instruments in support of biodiversity conservation, rural development or sustainable timber supply? What are the direct and indirect effects regionally and/or for future generations?

Discussions and policy deliberations about future alternatives should cover the whole knowledge chain affecting the development and utilisation of future forest resources: how different policies affect the forest management through the decision-making of forest owners; how different forest management regimes influence the forests and their development; and what are the socio-economic and ecological impacts of alternative forest management practices? Only then can balanced choices be made.

REFERENCES

Anttonen, T. and Petrov, A.P. (eds.) 1997. Potential of the Russian forests and forest industries. Research Notes 61. University of Joensuu, Faculty of Forestry. Finland. 144 p.

Buongiorno, J. and Gilless, J.K. 1987. Forest management and economics. Biological Resource management Series. University of California, Berkeley. MacMillan Publishing Company. NY, USA.

COMTRADE 1998. Global trade in wood products. United Nations Trade division. New York.

Csoka, P. 1998. Forest policy activities in the countries in transition in their preparation for the EU. In: Glück, P., Kupka, I. and Tikkanen, I. (eds.), Forest policy in the countries with economies in transition – ready for the European Union? EFI Proceedings 21. European Forest Institute. Pp. 9–20.

Diaci, J. (ed.) 1999. Virgin forests and forest reserves in Central and East European countries. COST E4. Forest Reserves Research Network. Proceedings of the meeting held in Ljubljana, Slovenia. Dept. of Forestry and Renewable Forest Resources. 171 p.

Dielen, L.J.M., Guegan, S., Lacour, P.-A., Mäki, P.K., Stolp, J.A.N. and Rytkönen, A. 2000. EU Energy policy impacts on the forest based industry – summary. Stichting Bos en Hout and AFOCEL. Wageningen, The Netherlands. 19 p.

EFISCEN's European Forest Resource Database, EEFR, http://www.efi.fi/projects/eefr/

European Commission 1997. Energy for the future: renewable sources of energy. COM 97:599. Brussels, Belgium.

Farrell, E.P., Fuhrer, E., Ryan, D., Andersson, F., Huttl, R. and Piussi, P. 2000. European forest ecosystems: building the future on the legacy of the past. Forest Ecology and Management 132: 5–20.

Fischler, F. 1998. Communication from the Commission to the Council and the European Parliament on a Forestry Strategy for the European Union. COM 1998/649. 25 p.

Glück, P., Oesten, G., Schanz, H. and Volz, K.-R. (eds.) 1999. Formulation and implementation of national forest programmes. Vol. I: Theoretical aspects. EFI Proceedings 30. European Forest Institute. 296 p.

Guo, B. and Bouton, Y.N. Unpublished. Application of a model for uneven-aged forest management to European forest resource projections – A test for the French uneven-aged forests.

Harmon, M.E., Franklin, J.F., Swanson, F.J., Sollins, P., Gregroy, S.V. et al. 1986. Ecology of coarse woody debris in temperate ecosystems. Advances in Ecological research 15: 133–302.

Haynes, R., Adams, D., Alig, R., Brooks, D., Durbak, I., Howard, J., Ince, P., McKeever, D., Mills, J., Skog, K. and Zhou, X. 2000. Projections of the US timber supply and demand – situation to 2050. USDA Forest Service. Draft results on the RPA Timber Assessment home page, http://www.fs.fed.us/pnw/sev/rpa/

Hees, A. van and Clerkx, S. 1999. Dead wood in forest reserves in the Netherlands. De Levende Natuur 100: 168-172. (In Dutch)
Kauppi, P.E., Mielikäinen, K. and Kuusela, K. 1992. Biomass and carbon budget of European forests, 1971 to 1990. Science 256: 70-74.
Kellomäki, S., Mohren, G.M.J., Karjalainen, T. and Lapveteläinen, T. 2000. Expert assessments on the likely impacts of climate change on forests and forestry in Europe. EFI Proceedings 34. European Forest Institute. 120 p.
Kennedy, J.J., Dombeck, M.P. and Koch, N.E. 1998. Values, beliefs and management of public forests in the Western world at the close of the twentieth century. Unasylva 49: 16-26.
Kirby, K.J. and Watkins, C. 1998. The ecological history of European forests. CAB International. Wallingford, UK. 373 p.
Kirschbaum, M.U.F., Fischlin, A., Cannell, M.G.R., Cruz, R.V.O., Galinski, W., Alvarez, A., Odera, J.A. and Xu, D. 1996. Chapter 1: Climate change impacts on forests. IPCC 1995 Assessment. Cambridge. Pp. 95-129.
Klimo, E., Hager, H. and Kulhavy, J. 2000. Spruce Monocultures in Central Europe – Problems and Prospects. EFI Proceedings 33. European Forest Institute. 208 p.
Köhl, M. and Päivinen, R. 1997. Study on European forestry information and communication system. 2 volumes. Luxembourg.
Kramer, K. and Mohren, G.M.J. 2001. Long-term regional effects of climate change on European forests: impact assessment and implications for carbon budgets. Final report of LTEEF II. Report 194. ALTERRA, Wageningen. Wageningen, ALTERRA.
Kuusela, K. 1994. Forest Resources in Europe 1950-1990. EFI Research Report 1. Cambridge University Press. 154 p.
Laitat, É., Karjalainen, T. Loustau, D. and Linder, M. 2000. Towards an integrated scientific approach for carbon accounting in forestry. COST E21 Workshop. Contribution of forests and forestry to mitigate greenhouse effects. Biotechnologie, Agronomie, Société et Environnement 4(4): 241-251.
Liski, J., Karjalainen, T., Pussinen, A., Nabuurs, G.J. and Kauppi, P. 2000. Carbon sinks and sources in trees of the European Union. Environmental Science and Policy 3: 91-97.
Martin, P.H., Valentini, R., Kennedy, P. and Folving, S. 1998. New estimate of the carbon sink strength of EU forests integrating flux measurements, field surveys and space observations: 0.17-0.35 Gt (C). Ambio 27: 582-584.
Mather, A.S. 1990. Global forest resources. Chapter 3. Historical perspectives on forest resource use. Timber Press. Portland, OR. Pp. 30-57.
Nabuurs, G.J. 2001. European forests in the 21st century: impacts of nature-oriented forest management assessed with a large scale scenario model. University of Joensuu Research Notes 130.
Nabuurs, G.J. and Päivinen, R. 1996. Large scale forestry scenario model – a compilation and review. EFI Working paper 10. European Forest Institute. 174 p.
Nabuurs, G.J., Päivinen, R., Sikkema, R. and Mohren, G.M.J. 1997. The role of European forests in the global carbon cycle – a review. Biomass and Bioenergy 13: 345-358.
Nabuurs, G.J., Pajuoja, H., Kuusela, K. and Päivinen, R. 1998. Forest resource scenario methodologies for Europe. EFI Discussion paper 5. European Forest Institute. 30 p.
Nabuurs, G.J., Schelhaas, M.J. and Pussinen, A. 2000. Validation of the European Forest Information Scenario Model (EFISCEN) and a Projection of Finnish Forests. Silva Fennica 34: 167-179.

Nabuurs, G.J., Pussinen, A., Karjalainen, T., Erhard, M. and Kramer, K. 2002. Stemwood volume increment changes in European forests due to climate change – a simulation study with the EFISCEN model. Global Change Biology 8: 304–316.

Nilsson, S., Sallnäs, O. and Duinker, P. 1992. Future forest resources of Western and Eastern Europe. International Institute for Applied Systems Analysis. The Parthenon Publishing Group. England. 496 p.

Niskanen, A. Pers. Comm. Unpublished results of a questionnaire done in the MOSEFA project. Reported values per country varied; some had reported gross income, some income after taxes, some included subsidies. European Forest Institute.

Nuutinen, T. and Kellomäki, S. 2001. A comparison of three modelling approaches for large scale forest scenario analysis in Finland. Silva Fennica 35(3): 299–308.

Päivinen, R., Lin, C., Ottitsch, A., Schuck, A. and Moiseyev, A. 1999a. Global overview of the European forests. In: Pelkonen, P., Pitkänen, A., Schmidt, P., Oesten, G., Piussi, P. and Rojas, E. (eds.), Forestry in changing societies in Europe. Vol. 1. University of Joensuu Press. Joensuu, Finland. Pp. 5–38.

Päivinen, R., Nabuurs, G.J., Lioubimov, A.V. and Kuusela, K. 1999b. The state, utilisation and possible future developments of Leningrad region forests. EFI Working Paper 18. European Forest Institute. 59 p.

Pajuoja, H. 1995. The outlook for the European Forest Resources and roundwood supply. Geneva, Timber and Forest Discussion Papers. ETTS V Working Paper. UN-ECE/FAO ECE/TIM/DP/4. Geneva. 59 p.

Parviainen, J., Little, D., Doyle, M., O'Sullivan, A., Kettunen, M. and Korhonen, M. 1999. Research in Forest Reserves and Natural Forests in European Countries – Country Reports for the COST Action E4: Forest Reserves Research Network. EFI Proceedings 16. European Forest Institute. 304 p.

Peck, T. 1998. General introduction to the main forest challenges worldwide: a European perspective. Proceedings of the 2nd International Forest Policy Forum (CTFC). Vol. 4. Solsona. Pp. 73–92.

Pelkonen, P., Pitkänen, A., Schmidt, P., Oesten, G., Piussi, P. and Rojas, E. (eds.) 1999. Forestry in changing societies in Europe. Vol. 1 & 2. University of Joensuu Press. Joensuu, Finland.

Pisarenko, A.I., Strakhov V.V., Päivinen, R., Kuusela, K., Dyakun, F.A. and Sdobnova, V.V. 2000. Development of forest resources in the European part of the Russian Federation. EFI Research report 11. Brill: Leiden, Boston, Köln. 102 p.

Pussinen, A., Schelhaas, M.J. Verkaik, E. Heikkinen, E., Päivinen, R. and Nabuurs, G.J. 2001. Manual of the EFISCEN model. Internal Report 5. European Forest Institute. 58 p.

Sallnäs, O. 1990. A matrix growth model of the Swedish forest. Studia Forestalia Suecica. No 183. Swedish University of Agricultural Sciences. Faculty of Forestry. Uppsala. 23 p.

Schelhaas, M.J., Nabuurs, G.J., Sonntag, M. and Pussinen, A. 2002. Adding natural disturbances to a large scale forest scenario model and a case study for Switzerland. Forest Ecology and Management 167: 13–26.

Schelhaas, M.J., Varis, S., Schuck, A. and Nabuurs, G.J., 1999. EFISCEN's European Forest Resource Database, European Forest Institute, Joensuu, Finland, http://www.efi.fi/projects/eefr/

Solberg, B. 1995. Forest biomass as carbon sink – economic value and forest management and policy implications. Journal of Environmental and Resource Economics 5(10): ???.

Spiecker, H., Mielikäinen, K., Köhl, M. and Skovsgaard, J.P. 1996. Growth trends in European forests. EFI Research Report 5. Springer-Verlag, Heidelberg, Germany. 372 p.

Trømborg, E., Buongiorno, J. and Solberg, B. 2000. The global timber market: implications of changes in economic growth, timber supply, and technological trends. Forest Policy and Economics 1: 53–69.

UN-ECE and European Commission 1999. Forest condition in Europe. Results of the 1998 survey. Convention on Long-Range Transboundary air pollution. Brussels, Belgium. 43 p.

UN-ECE/FAO 1992. The Forest Resources of the Temperate Zones, the UN-ECE/FAO 1990 Forest Resource Assessment, General Forest Information. Geneva, Switzerland. 347 p.

UN-ECE/FAO 1996. European timber trends and prospects: into the 21st century. Geneva, Timber and Forest Study Papers No 11. ECE/TIM/SP/11. United Nations New York, Geneva. 103 p.

UN-ECE/FAO 2000. Forest Resources of Europe, CIS, North America, Australia, Japan and New Zealand. Geneva, Timber and Forest Study Papers, No 17. 445 p.

Uuttera, J., Maltamo, M. and Hotanen, J.P. 1997. The structure of forest stands in virgin and managed peatlands: a comparison between Finnish and Russian Karelia. Forest Ecology and Management 96: 125–138.

Valentini, R., Matteuci, G. and Dolman, A.J. 2000. Respiration as the main determinant of carbon balance in European forests. Nature 404: 861–865.

Watson, R.T., Noble, I.R., Bolin, B., Ravindranath, N.H., Verardo, D.J. and Dokken, D.J. 2000. Land Use, Land Use Change and Forestry, a Special Report of the IPCC. Intergovernmental Panel on Climate Change, Cambridge University Press. Cambridge, UK. 377 p.

Wermann, E. 1999. Accelerating tree growth – policy consequences. In: Karjalainen, T., Spiecker, H. and Laroussinie, O. (eds.), Causes and consequences of accelerated tree growth in European Forests. EFI Proceedings 27. European Forest Institute. Pp. 9–14.